Translation, Reception and Canonization of *The Art of War*

The Art of War by Sun Tzu is an ancient yet invaluable Chinese military classic that is still relevant today. This book presents a systematic and in-depth investigation into the translation and reception of *The Art of War* in Western strategic culture.

Aided by three self-built corpora, this book adopts a mixed method of qualitative and quantitative analysis, and takes both the core text and paratexts of *The Art of War* into consideration. The author highlights the significance of proper approaches to translating culture in regards to the core text and effective measures of culture reconstruction in regards to the paratexts. It is revealed by this investigation that the translated Sun Tzu has undergone three major stages before its canonization in Western discourse. The findings bring into light the multiple factors that contribute to the incorporation of Sun Tzu's strategic wisdom into Western culture.

For scholars interested in translation studies, (critical) discourse analysis, as well as strategic studies, this book provides fresh insights and new perspectives.

Tian Luo is Professor of translation studies at Chongqing Jiaotong University. He received his PhD Degree in Linguistics (English) at the University of Macau in 2017. His research interests fall into the fields of military translation history, discourse analysis approach to translation studies and corpus-based translation studies. His publications include several monographs, textbooks in English-Chinese translation and more than 30 articles in journals such as *Perspective*, *Babel* and *LANS-TTS*. He is currently leading a research project sponsored by the National Social Science Foundation of China after the completion of three other projects.

Routledge Advances in Translation and Interpreting Studies

For more information about this series, please visit www.routledge.com/
Routledge-Advances-in-Translation-and-Interpreting-Studies/book-series/
RTS

Translation, Reception and Canonization of *The Art of War*

Reviving Ancient Chinese Strategic Culture

Tian Luo

Routledge
Taylor & Francis Group

LONDON AND NEW YORK

This book is published with support from the National Social Science
Foundation of China under Grant Number 17BYY200

First published 2022
by Routledge
4 Park Square, Milton Park, Abingdon, Oxon OX14 4RN

and by Routledge
605 Third Avenue, New York, NY 10158

Routledge is an imprint of the Taylor & Francis Group, an informa business

British Library Cataloguing in Publication Data
A catalogue record for this book is available from the British Library

Library of Congress Cataloging in Publication Data
A catalog record for this book has been requested

ISBN: 978-0-367-45846-1 (hbk)
ISBN: 978-1-032-24533-1 (pbk)
ISBN: 978-1-003-02572-6 (ebk)

DOI: 10.4324/9781003025726

Typeset in Times New Roman
by Newgen Publishing UK

Contents

Figures

Tables

Abbreviations

ST	source text
TT	target text
SL	source language
TL	target language
SC	source culture
TC	target culture
CDA	Critical Discourse Analysis
UK	The United Kingdom
US	The United States
AOW	The Art of War

Acknowledgements

This book is based on my PhD Thesis finished in 2017. Without the constant support from many distinguished scholars, helpful colleagues and caring relatives, this book would have been an impossible mission for me. I would like to take this chance to express my sincere gratitude to all of them.

My gratitude goes first and foremost to Professor Meifang Zhang, my PhD supervisor, who is always enthusiastic about translation research, efficient at research project management, young at heart and elegant in every movement. All through the years I spent on the beautiful campus of the University of Macau, I have been benefited from her mentoring. She led us to international conferences and enlightened us with rich lessons in translation theories and research methodology, especially the functional and discursive approach to translation. Being a rigorous scholar, she sharpened my skills in critical thinking and academic writing, guided my research in every aspect, directed me to the right track while I was preparing my PhD thesis, and reviewed it with great patience and meticulous efforts. Her passion for academic excellence and great care and love for the people around her is a ray of sunshine that will always light my road in my future academic career.

I owe my great gratitude to the members of my PhD thesis advisory committee and oral defense committee, among whom are Prof. Defeng Li, Prof. Yuanjian He, Prof. Yifeng Sun, Prof. Andrew Moody and Prof. Christopher Kelen. They generously shared with me their expertise, advice and insightful questions that drove me to deep contemplation and helped me come up with a satisfactory work.

I would also like to thank other teachers at the University of Macau who had taught and inspired me: Prof. Martin Montgomery, Prof. John Corbett, Dr. Victoria Lei, Dr. James Li and Dr. Hari Vankatesan.

I am thankful for my supporting and lovely colleagues, with whom I have had a wonderful time on campus: Dr. Hanting Pan, Dr. Xi Chen, Dr. Bingjian Qin, Dr. Hongqiang Zhu, Dr. Xiaoping Wu, Dr. Sijing Zhou and Dr. Chengcheng You, among many others. I always looked up to the senior PhD colleagues: Prof. Li Pan, Dr. Qianhua Ouyang, and Dr. Hong Qian from whom I have learned a lot.

To those scholars and friends elsewhere who have offered valuable advice and aid in various forms, I would like to express my thanks: Prof. Juliane House, Prof. Jeremy Munday, Prof. Kirsten Malmkjær, Prof. Xuanming Luo, Prof. Lei Mu, Prof. Anjiang Hu, Prof. Jianping Huang, Prof. Xiangbin Wang, Prof. Ziman Han, and Director Mingzhi Chen and many more. I would also like to thank Prof. Qiyi Liao, Prof. Sihui Mao and Prof. Lu Shao for their constant support in particular.

Many thanks go to the team of experts who assisted the publication of this book: Katie Peace, the senior editor and publisher at Routledge, for all her help, advice and encouragement over many months it took to complete this book; James Bradshaw, who with a pair of sharp eyes, provided expert proof reading and editing service; Sunantha Ramamoorthy, the project manager for my book publication and many others whose names I have not yet known.

I thank Taylor & Francis for allowing me to use some content with small changes from a published journal paper: Tian Luo & Meifang Zhang. 2018. "Reconstructing cultural identity via paratexts: A case study on Lionel Giles' translation of *The Art of War*", *Perspectives* 26 (4): 593–611.

Last but not least, my deep appreciation is delivered to my dear mother whose understanding and support has been vital to my study. And I hold gratitude to my deceased father whom I miss so much and who taught me by example that it pays back to be diligent. Special thanks go to my wife Yi Li for her love and tremendous assistance during my study, to my daughter Zidan Luo, whose laughter is the main source of reassurance when I am in times of difficulty. Also, I would like to thank my parents-in-law, my elder sister and younger brother for their support.

This monograph is also supported by the National Social Science Foundation of China under Grant Number 17BYY200.

To those who love me and whom I love, this book is sincerely dedicated.

1 Introduction to *The Art of War* and its English translations

The Art of War by Sun Tzu is a timeless Chinese military masterpiece with worldwide influence. It enjoys an important status in Chinese strategic culture and boasts multiple translations in many languages. The main purpose of this study to find out how *The Art of War* is translated into English, to what extent it is received in the Western discourse and why it is received in the West as it is.

1.1 *The Art of War,* a key text of Chinese strategic culture

The Art of War (hereinafter referred to as *AOW*) constitutes a rather valuable topic in translation studies since it is a Chinese military classic as well as a key text of Chinese strategic culture and, with its multiple translations into many other languages, it exerts a far-reaching influence on military and strategic thinking around the globe.

AOW (孙子兵法 in Chinese ideographic characters, and *Sunzi Bingfa* in the modern Chinese Pinyin system), the earliest military thesis known in human history, is believed to have been authored by Sun Wu(孙武), a legendary general who lived at the end of Spring and Autumn period, roughly fifth century BC, in northern China. Sun Wu was addressed respectfully as Sun Zi (孙子), since Zi (子) in ancient Chinese was an honorific title that meant a "master". In English, this mysterious general is referred to as "Sun Tzu" in the old-fashioned Wade-Giles Romanization system or "Sun Zi" in the modern Chinese Pinyin romanization system.

Sun Tzu lived in a time of great turmoil in China, as many feudal vassal states competed for power and control by means of military force and stratagems. His profound knowledge and skills as a wise warrior consequently found their expression in the concise book he composed. According to *Shiji* (*Records of the Grand Historian*) by Sima Qian in the first century BC, Sun Wu was a native of the State of Qi who moved to the southeastern kingdom of Wu. He impressed King Helü of Wu (about 514–495 BC) with his outstanding ability to train even the king's dainty concubines into an army of properly drilled and well-disciplined soldiers. Having been appointed as a general, Sun Tzu led many victorious battles with his strategic genius in the campaigns against the states of Chu, Qi and Yin. His fame spread among the feudal princes.

DOI: 10.4324/9781003025726-1

With a history of more than 2,500 years, *AOW* is also "the most brilliant and widely applied strategic book ever written" (Huang 2008, 15). Composed of 13 chapters in slightly more than 6,000 Chinese characters, it features a systematic argumentation of strategic issues, including views on war, military leadership, strategy and tactics, terrain and espionage.

The first three chapters of *AOW* address the issue of warfare as a national concern from a macro perspective. The first chapter, "Laying Plans", stresses the importance of warfare to a nation and discusses the five fundamental factors of war (the moral law, weather, terrain, leadership and management) and the seven elements that determine the outcomes of military events. Sun Tzu articulates the role of the commander's calculation in advance for greater chances of victory and emphasizes the role of deception in warfare. The second chapter, "Waging War", explains the economy of warfare. Sun Tzu values the quick, effective and successful engagement that is won with minimal costs of national resources. Chapter 3, "Attack by Stratagem", highlights the central goals of warfare—victory and peace. It argues that the best-fought wars are won by attacking an enemy's strategy, and the costly and time-consuming siege warfare should be a last resort. Sun Tzu also emphasizes foreknowledge of oneself and the foe before fighting and lists five factors that lead to a victory: knowing when to fight, knowing how to deploy strength, knowing how to motivate troops, good preparation for the unexpected and having a ruler who does not meddle with a talented commander.

The next three chapters of *AOW* provide the basic rules of military strategy. Chapter 4, "Military Disposition", stresses again that a general shall estimate and weigh the strength of the enemy and prepare well in order to assure victory. Sun Tzu describes his approach to war in terms of opposing forces that complement and give rise to each other: vulnerable/invulnerable, defense/attack and lack/plenty. The fifth chapter, "Strategic Military Power", explains the necessity of managing one's troops in all situations and the use of creativity and timing to build up an army's momentum or strategic advantage. Sun Tzu attaches great importance to the employment of indirect, unorthodox and unexpected forces. Chapter 6, "Weaknesses and Strengths", reveals how a general should respond to the ever-changing battlefield environment to secure opportunities of victory, which result from the relative weakness of the enemy.

Chapters 7 and 8 delve into the details of maneuvering an army and of fighting tactics. Chapter 7, "Military Maneuvers", explains proper methods for maneuvering one's troops in different situations. It pays special attention to the use of signals, such as banners and drums, and cautions against overloading soldiers. Chapter 8, "Variation of Tactics", is short but gives an important suggestion: vary tactics to confuse the enemy. It illustrates the significance of flexibility in an army's responses to shifting circumstances.

The next three chapters are all related to the issue of landscape and terrain. Chapter 9, "Movement and Development of Troops", focuses on evaluating the intentions of the enemy. It stresses that a commander should think like his enemy and identify his tactics and he should know approaches to marching armies through different landscapes. Chapter 10, "Terrain", regarding the degree of accessibility and danger, classifies six different types of ground positions that a commander shall ponder, each with a variety of pros and cons. It also warns against six calamities that lead to the collapse of an army. Chapter 11, "The Nine Battlegrounds", is the longest one, and describes nine common battlefields in a competitive campaign which call for the commander's acute recognition and appropriate response.

The last two chapters are even more specific, providing detailed information on fire attacks and the use of spies. The 12th chapter, "Attacking with Fire", systematically examines five targets for fire attack and appropriate responses to such attacks. This chapter discusses the use of natural resources aimed at the destruction of an opponent. The 13th chapter, "Espionage", expounds the five types of secret agents and provides suggestions on how to manage different spies.

In addition to his military insight, Sun Tzu was a genius for epigrams. *AOW* is rich in concise, profound and well-balanced sentences that are crystal clear and memorable. Many sentences from *AOW* have become celebrated dictum in China, such as "All warfare is based on deception", "Just as water retains no constant shape, so in warfare there are no constant conditions" and "To fight and conquer in all your battles is not supreme excellence; supreme excellence consists in breaking the enemy's resistance without fighting".

With its systematic inquiry of warfare in pithy yet aphoristic language, *AOW* stands out as a military masterpiece with an enduring appeal for Chinese readers. Ever since its composition, it has been favorably commented, scrupulously researched and extensively applied by a rather long list of distinguished Chinese military generals, scholars and politicians. From the Warring State period (475 BC–221 BC), *AOW* began to reach other parts of China and became increasingly popular, providing guidance for military conflicts as well as non-military competitions in diplomatic, management and medical contexts (Anonymous, 2015). Such popularity was evidenced by a large quantity of imitative remarks or quotations from Sun Tzu in books such as *Wu Zi (Master Wu)*, *Wei Liao Zi (Master Wei Liao)*, *Liu Tao (The Six Principles of War)*, and *Sun Bin's The Art of War*. Ts'ao Ts'ao, a strategist and politician in the early third century, furnished the earliest known commentary to *AOW*. Ts'ao acclaimed "Many books have I read on the subject of war and fighting; but the work composed by Sun Tzu is the profoundest of them all" (quoted in Giles 2007, xxxv). In the following centuries, *AOW* was interpreted, commented, annotated or edited by hundreds of ancient and modern scholars, among them Li Ch'üan (eighth century), Tu Mu (803–852), Mei Yaoch'en (1002–1060), Sun Hsingyen (1753–1818), Guo Huaruo (1904–1955), Li Yuri (1908–1955) and Li Lin (1948–). Their contributions have brought about plenty of *AOW* editions. In addition to commentators and editors, a school of generals perused the pages of *AOW* with enthusiasm and applied its lessons on the battlefield. Among these military strategists and political leaders are Han Xin (231 BC–196 BC), Lű Meng (178–220), Yue Fei (1103–1142), Yu Dayou (1503–1579), Qi Jiguang (1528–1588) and Mao Tse-tung (1893–1976), only to name a few. For instance, Mao, inspired by Sun Tzu's maxim "Avoid the enemy when he is full of vigour, strike when he is fatigued and withdraws" (避其锐气，击其惰归), proposed his well-reputed mantra of guerrilla war: "The enemy advances, we retreat; the enemy camps, we harass; the enemy tires, we attack; the enemy retreats, we pursue" (敌进我退，敌驻我扰，敌疲我打，敌退我追).

AOW was also listed among the textbooks for many military personnel. It became the leading text in an anthology entitled "*The Seven Military Classics*" issued by Emperor Shenzong of the Song Dynasty in 1080. That anthology paved the way for Sun Tzu's greater popularity. In the Ming and Qing Dynasties, *AOW* was included in the royal martial examination for the selection of military talents, which further urged more students to ponder it. Sun Tzu is still taught in contemporary Chinese universities for servicemen.

Gradually, *AOW* became the most prominent and valuable asset in the repository of Chinese strategic culture. In a general sense, military classics are a key component of strategic culture, which is an essential issue for mankind since it concerns war and peace. It is a consensus among Chinese scholars, particularly military experts such as Wu (2004b), Wang (2004) and Li (2005), that *AOW* contains the core values of ancient Chinese strategic culture. Many foreign experts also agree, as Howard (2013, 18) puts it: "Perhaps the biggest influence on Chinese military strategy, Sun Tzu's *The Art of War*, has been and remains a seminal philosophy underpinning Chinese strategic culture."

Among multiple definitions of strategic culture proposed by a number of scholars such as Snyder (1977, 8), Booth (1979, 121), Johnston (1995, 46), Gray (1999, 131–133) and Scobell (1999, 479), the one proposed by Kerry Longhurst is comprehensive and typical:

> A strategic culture is a distinctive body of beliefs, attitudes and prac-tices regarding the use of force, which are held by a collective and arise gradually over time, through a unique protracted historical process. A strategic culture is persistent over time, tending to outlast the era of its original conception. It is not a permanent or static feature. Rather, a strategic culture is shaped by formative experiences and can alter, either fundamentally or piecemeal, at critical junctures in that collective's experiences.
>
> (Longhurst 2004, 17)

Strategic culture usually consists of three parts: foundational elements, regulatory practices and security policy standpoints. Specifically, founda-tional elements "comprise basic beliefs regarding the use of force that give a strategic culture its core characteristics", which are highly resistant to change (Longhurst 2004, 17–18). On the behavioral side, the regulatory practices "actively relate and apply the substance of the strategic culture's core to the external environment" (Longhurst 2004, 17–18). Security policy standpoints are "the contemporary, widely accepted interpretations as to how best core values are to be promoted through policy channels, in the sense that they set the preferences for policy choices" (Longhurst 2004, 17–18).

Strategic culture is "collective", "unique" and "historical", and features both persistence over time and the possibility for change (Longhurst 2004, 18). It is also characterized by its dynamism, and "the existence and functioning of the three components mean that a strategic culture is in a continual state of self-evaluation" (Longhurst 2004, 18).

Strategy and culture are interrelated. On one hand, strategic thinking is shaped and conditioned by the overall cultural context; on the other hand, culture is enriched by the development of strategy. As Lieutenant General Li Jijun, the former vice president of the Chinese Academy of Military Sciences remarks:

> Culture is the root and foundation of strategy. Strategic thinking, in the process of its evolutionary history, flows into the mainstream of a country or a nation's strategic culture. Each country or nation's strategic culture cannot but bear the imprint of cultural traditions, which, in a subconscious and complex way, prescribes and defines strategy making.
>
> (Li 1997; quoted from Scobell 2002, 1)

The role and impact of strategic culture on war planning and security policymaking can never be overestimated. Strategic culture has become an integral part of the vocabulary of national security and international relations. Snyder (1977, 9) believes that strategic culture "guides and circumscribes thought on strategic questions, influences the way strategic issues are formulated, and sets the vocabulary and conceptual parameters of strategic debate". Booth (1990, 121) notices that it helps shape but not determine a nation's interaction with others in the security field, "helps shape behavior on such issues as the use of force in international politics, sensitivity to external dangers, civil–military relations and strategic doctrine". In Gray's view (1999, 133, 142, 144), "strategic behavior cannot be beyond culture", and strategic culture "is a guide to action" which "finds expression in distinctively patterned styles of strategic behavior". Longhurst (2004, 21) posits that a strategic culture will "constrain behaviour by excluding certain options and facilitate behaviour in various intensities".

The importance of *AOW*, therefore, lies not only in its in-depth and comprehensive analysis of military affairs, but also in its leading role in defining and representing the Chinese strategic culture and providing guidance for the use of military force over the course of time; not only in its outline of ancient strategic thinking, but also in its contemporary relevance. This explains the importance for a better understanding and systematic analysis of military classics such as *AOW*.

1.2 Translations of *The Art of War*: a brief history

Classics inevitably beget (re)translations. Classic texts are usually the hallmark of a certain culture and the asset of a given community, society or nation. Thanks to their unique and enduring quality, they excel and stand out from a school of competitors in fields such as literature, religion, politics, medicine, business and military strategy. They become well established in their source language community for a rather long time. Usually, they do not end their journeys within their own linguistic or cultural community. Instead, they spread "across spatial and temporal boundaries, even to those whose language and culture are different" (Lianeri and Zajko 2008, 5). With each journey propelled by translation, a classic work is most likely bestowed a chance for new life.

As Blanchot (1983, 84) highlights, "[c]lassical masterpieces live only in translation". On one hand, translation helps expand the geographic scope of a classic text, contributing to its dissemination among a wider audience. On the other hand, the act of translating suggests "a form of negotiation of the gap between past and present meaning" (Lianeri and Zajko 2008, 9), which brings about the contemporary relevance of the classic. With both geographic and chronological boundaries crossed, the enduring quality of a masterpiece is interpreted anew in translation, perceived and finally passed on by successive generations of readers in other cultures. Venuti (2008, 28) stresses that "translation functions as one cultural practice through which a foreign text attains the status of a classic".

Among the classics on varied topics, military masterpieces are extremely important since they address the issue of war and peace, an eternal theme of humanity, where life and death are at stake. Translation of the military classics of the East and West has been phenomenal because it started as soon as the two civilizations encountered each other and continued with increasing numbers. Translated military classics are assumed to have a huge impact on the way military conflicts are reviewed and researched, conducted and ended.

Naturally, *AOW* boasts a large number of translations in Asian languages and remains the most influential strategy text in East Asian warfare. Around 516 A.D., *AOW* found its way into Japan via Korea (Griffith 1963, 169). The first Japanese translation appeared as early as the year 1660 (Wu 2004a, 92) and there are currently at least 230 different Japanese versions issued in Japan (Su and Azusen 2009). In Korea, more than 220 editions of Sun Tzu in the Korean language have been published (Shao 2013, 98).

With multiple translations and interpretations, *AOW* has gradually been bestowed the status of "the most important military treatise in Asia", according to the historian and translator Ralph D. Sawyer (1993, 149). During the Sengoku period (c. 1467–1568), the Japanese general Takeda Shingen (1521–1573) is said to have become almost invincible in all battles because he practiced what he learnt from *AOW*. He was inspired by the book to set up the famous battle standard "Fūrinkazan" (Wind, Forest, Fire and Mountain), meaning to deploy an army as fast as the wind, silent as a forest, ferocious as fire and immovable as a mountain.

During the Vietnam War, some Vietcong officers extensively studied *AOW* and reportedly could recite entire passages of *AOW* from memory. General Võ Nguyên Giáp, the main PVA military commander in the Vietnam War, was an avid student and practitioner of Sun Tzu's ideas. He successfully implemented tactics described in *AOW* during the Battle of Dien Bien Phu, ending major French involvement in Indochina and leading to the accords which partitioned Vietnam into North and South.

Since the eighteenth century, readers have witnessed the emergence of many *AOW* translations into Western languages, which helped *AOW* further spread its influence around the globe. The first Western version of *AOW* was rendered into French by Jesuit Jean Joseph Marie Amiot in 1772 (re-published in 1782). *AOW* has now been translated into almost every major Western language and printed or reprinted in various editions, widely available to Western readers. According to statistics from the end of 2011, *AOW* translation could be found in more than 30 foreign languages, including French, Russian, German, Czech and English. Translations of *AOW* have been published in over 34 countries, including Korea, Japan, France, Russia, Great Britain, and the United States (Su 2011, 154). More often than not, there exist multiple versions in each foreign language by different translators. These different editions may help exert the influence of *AOW* on the global strategic culture.

Specifically, *AOW* boasts multiple English translations over a span of more than 110 years, with the earliest one, by Everard Calthrop, published in 1905 (Calthrop 1905) and the most recent one in 2020 (Nylan 2020). According to my survey, at least 60 translators have contributed their English versions (see Appendix 1: A list of English translations of *The Art of War*). Moreover, these English translations have been published in as many as 200 editions (Su 2011, 149), available in not only hard copies but digital forms. As a matter of fact, it would be a very demanding job now to count the number of editions of English *AOW* around the globe.

The first English translation was attempted by British officer Everard Ferguson Calthrop in 1905 under the title *The Book of War*, and was revised in 1908. Unsatisfied with Calthrop's 1905 translation, which was based on a Japanese source text and branded with Japanese cultural identity, Lionel Giles, the British sinologist, came up with the first annotated English translation in 1910. Giles' (1910) translation was complete and bilingual, including both English and Chinese texts printed in parallel, and more importantly, was supplemented with a large number of notes to illustrate Sun Tzu's military thoughts expressed in pithy Chinese. Giles' scholarly translation was a great success, catching the attention of strategists and ordinary readers and being repeatedly printed.

English (re)translations of *AOW* have been finished by translators with various identities from different social backgrounds in different disciplines. Some were by military officers and experts (e.g. Calthrop 1905; Griffith

1963; Li 1938), some by sinologists (e.g. Giles 1910; Minford 2002; Sadler 1944), some by philosophers (e.g. Ames 1993; Sawyer 1993), others by professors and executives in business and management (Chen and Chan 1998; Gagliardi 1999) and still others by diplomats (Yuan, Shibing 1987/1990). Most translators work independently and individually (Wing 1988), while others cooperate closely as a group (The Denma Translation Group 2001). More than 20 translators to English are of Chinese origin (Lin, Wusun 2003), while others are from different nationalities and origins (Clements 2012). An overwhelming majority of the translations were done by males, while only a recent translation was finished by a female (Nylan 2020).

With different approaches adopted by the translators, these English translations exist in various forms and have distinct features. For instance, most translations are monolingual, represented in English only, while several are bilingual, including ancient Chinese text as well as its English version (e.g. Ames 1993); still others also provide modern Chinese interpretation (Xiao 2013) or Pinyin spellings (e.g. Richter 2004). Some versions only translate the core text (e.g. Chohan and Bellenteen 2003), while many others are rich in paratexts, such as prefaces, forewords, annotations and appendices (e.g. Cleary 1998). Most versions are monomodal presented only in verbal signs, while some are multimodal, rich in pictures, graphs and figures (e.g. Hagy 2015; Sui 2004). Most translations are literal renditions of the source text in a sentence-to-sentence fashion, while a few are a re-organized version of the source text(e.g. Tang 1969), excerpted (Luo 1995) or extensively adapted (Michaelson and Michaelson 2010).

Such a great variety of translations, either in English or other languages, has helped *AOW* reach different types of audiences around the globe. Since English serves as a lingua franca in the contemporary world, those translations of *AOW* in English deserve special attention if we are to conduct a case study of military classic translations and their influences in the West.

1.3 Previous research on the English translation of *AOW*

As was mentioned in Section 1.1, strategic culture plays a rather essential part in the making of contemporary strategy of a nation, and the importance of translation of military classics for strategic culture cannot be overemphasized. The translation of *AOW*, a key text of strategic culture, for the English-speaking world may call for earnest academic inquiry. This section, therefore, tries to provide a literature review of the research on the English translations of AOW.

It can be summarized that the English translations of *AOW* have received enthusiastic attention from the academic community, especially from Chinese scholars. Generally speaking, three schools of scholars—military history, comparative literature and linguistic (translation) —have attended the issue of *AOW* translations. Many military historians have acknowledged that *AOW* translations have brought fresh insights to Anglo-American military theories, redefined and re-theorized the Western strategic thinking and eventually turned *AOW* into a canonical reading in international military culture (Yuen 2014; Wu 1996; Yu 2001). Some comparative literature scholars, such as Yang (2012, 2017), have listed and compared the English translations of some cultural terms and the comments by Western scholars.

Linguistic research on the English translations of *AOW* began as early as 1965, when D.C. Lau (1965) published his critical analysis of the linguistic choices in Lionel Giles' translation. Many scholars joined the effort to investigate English translations from various perspectives. According to the survey by Luo and Zhang (2015, 55), there are at least 160 academic texts on the translations of *AOW* by the year 2015, including 3 monographs, 111 journal articles, 3 PhD theses, 38 master theses and 5 book chapters, with the overwhelming majority published in China. Still, more linguistic scholars keep adding their contributions.

This large body of research literature by linguists and translation scholars involves 21 English versions, covers a variety of topics and features interdisciplinary perspectives, different approaches and methodologies. The topics in these publications include examinations of culture-specific items, style, translation approaches, the dissemination of translated texts, retranslations and a translator's subjectivity. For instance, a comparative etymological study, conducted on the ancient culture-loaded words in two English versions of *AOW*, found that the culture gap and the translators' lack of cultural awareness impaired the translating of the culture-loaded words (Huang 2013). In another paper, Samuel Griffith's (1963) and Lin Wusun's (2003) versions were compared under the framework of Foucault's power discourse theory and found that the translation of culture-specific items was affected by both individual and social power discourse (Wang 2011).

Many studies have been conducted from interdisciplinary perspectives, drawing on theories from cultural studies, sociology and philosophy. Almost half of the journal papers adopted cultural approaches, using a theory of polysystems, manipulation, discourse and power, or cultural memes (Luo and Zhang 2015, 57). For instance, Wu (2013), in his monograph, discussed seven translated versions of Sun Tzu from the intercultural communication perspective, applying a wide range of theories, such as social semiotics, synergetics, hermeneutics, cultural convergence theory, and communication accommodation theory. Zhang (2014) reflected on the competition strategy among the retranslations of *AOW* from the perspective of Misreading Theory to find out that the later translators tend to outwit their predecessors by misreading, criticizing, revising and rewriting the precursors' translations. Zhang and Li (2011) compared Yuan Shibin's translation with Lin Wusun's in light of the polysystem theory to discover that these two versions display different translation approaches and styles, with Yuan focusing more on acceptability of terminology, and Lin on adequacy. Their paper reveals the important role of patronage in affecting the choice of translation strategy. Huang (2009) was interested in the impact of ideology on the retranslation of classical works. He found that John Minford's misinterpretation of Sun Tzu's thoughts and the translation approach of vilification resulted from the influence of Orientalism and the translator's identity as an Orientalist. Song (2012), referring to Bourdieu's sociological concepts of cultural capital and field, found that in retranslating *AOW*, individual translators were inclined to use various kinds of cultural capital (embodied, objectified and institutionalized) to outmatch the competition not only within textual practice but also well beyond it.

Although there exists a large body of research on the translations of *AOW*, their inadequacies or weaknesses are still evident in the following aspects. First of all, the issue of how strategic culture is reconstructed in a target language and received in target culture has remained under-investigated. Compared with the great advances in the study on translation of other types of cultures (e.g. religious culture, with translations of the Bible, Buddhist scripts and the Koran), strategic culture in translation has remained an under-developed area for a rather long time. Few in-depth studies have been conducted on the role of translation in ushering one strategic text into another culture. Specifically, *AOW* is first and foremost a book on warfare and studies on its translation rarely devote enough attention to the issue of strategic culture. Although some studies have investigated the translation of cultures or conducted cultural studies of the translation of *AOW*—for instance Wei (2018) addressed the translation of ecological, material, social, religious and linguistic culture in *AOW* and Huang (2018) examined cultural values, misinterpretation, hybridity and capital in the English translation of *AOW*—they tend to ignore the key issue of strategic culture, which is the very essence of the military classic *AOW.* Certainly, no in-depth research has been conducted about how ancient Chinese strategic culture in *AOW* is reconstructed in translations and received in Western military discourse.

Second, a great gap exists between the relevant studies by military historians and comparative literature scholars and those by linguistic scholars. The perspective from which military historians investigated Sun Tzu's translation is rather broad and abstract and they tend to neglect linguistic nuances. The studies by linguistic (translation) scholars tend to focus on minute linguistic issues, while ignoring to some extent the macro social-historical background. Previous studies, therefore, have failed to establish a link between translation and the evolution of strategic culture, or produce concrete evidence for the role of translation in the global dissemination of Chinese strategic culture (especially *AOW*). Further investigation still needs to find out how translators deal with the military terms and principles and how translation contributes to the reception and circulation of strategic culture. It is also necessary for linguistic (translation) scholars to look at the translation of strategic culture from a larger social-cultural framework since no strategic issue can be confined within a narrow linguistic boundary.

Third, although a large volume of paratexts constitute an integral part of *AOW* (such as the commentary by Chinese scholars) and its translations, how the paratexts are translated and what role they play in the reception of *AOW* in the West has been neglected in most cases. Even though some scholars tried to approach these paratexts, they tended to do it in isolation and neglected to some extent the partnership or the close cooperation between the core text and paratexts.

Last but not least, very few studies have provided quantitative evidence of how *AOW* has been translated and received, while the overwhelming majority of previous studies have been qualitative, featuring example analysis. Particularly, there is almost no statistical and convincing analysis to reveal how these translations have been received. For instance, even though some previous research involves as many as 21 different English versions, no ample statistical evidence has been given to identify which are the most influential ones, and with what attitude they were received, positive or negative.

1.4 Research objectives, questions and significance

Although military conflicts occur around the world almost every day and related strategic works are translated globally, military classic and strategic culture has been inadequately heeded by translation scholars. As mentioned in Section 1.3, although previous studies on the translation of *AOW* are rich and display a rising trend, there are still some deficiencies to be remedied. Setting out upon such ground, this study aims to approach the topic of strategic culture and military classic translation through a case study of *AOW*, so as to find out how, why and to what effect the ancient Chinese military culture has been translated and received by the Western strategic culture. My special focus has been on English translations since English is a lingua franca that invites the largest number of readers.

There are three major objectives of this study: (1) to identify the approaches and methods in translating and reconstructing the military culture in TTs, taking into consideration the roles of both core texts and paratexts; (2) to map out the reception and impact of the translated Sun Tzu in the target culture; and (3) to explore possible social-cultural constraints over the translation and reception of *AOW* in the western military discourse.

In accordance with these objectives, some specific research questions are raised:

(1) What makes *AOW* a military classic? How does the core text, in partnership with its paratexts, contribute to the canonization of the ST in the Chinese culture?
(2) How is the strategic culture translated in the core texts and reconstructed in the paratexts in different English versions? Do the methods of translation and reconstruction have an impact on the reception of Sun Tzu?
(3) How is the translated Sun Tzu received in the target culture? Does the reception go through different stages and finally reach the status of canonization in the target culture?

(4) What is the social-cultural background of the translation and reception of *AOW* in the West? Do power and ideology of any kind effect the process of translation and reception?

(5) What are the implications of this case study of *AOW* for the translation of strategic culture and military classics?

This study finds it necessary to investigate how language barriers are overcome in these translations. It is also worthwhile to examine how these translations manage to reconstruct the Chinese strategic culture so as to find easy access into Western culture, and how these translations are received by the Western audience. Especially, the critical link between the translation methods and the reception of *AOW* in the West needs to be established so as to highlight the role of translation in the reconstruction of Chinese strategic culture.

First of all, this study examines the fundamental issue for the survival of *AOW* in a foreign culture: translation methods. It is no exaggeration to say that the future fate of classics lies in the hands of translators. If not properly handled, "a foreign text may well lose its native status as a classic and wind up not only unvalued, but unread and out of print" (Venuti 2008, 28). If wisely treated, the process of translation may increase the value of a classic by generating promotional devices and by enabling diverse modes of reception (Venuti 2008:28). Translation approaches are vital not only for the text's own status as a classic, but also for the culture it represents or embodies, as Lefevere points out:

> If a text is considered to embody the core values of a culture, if it functions as that culture's central text, translations of it will be scrutinized with the greatest of care, since "unacceptable" translations may well be seen to subvert the very basis of the culture itself.
>
> (Lefevere 1992, 70)

Second, this study also investigates the social-cultural context in which *AOW* translations are done, and tries to establish the link between the translation approach and reception. Translation approaches and methods are not the only factors onto which the status of a translated classic is hung. Looking beyond texts, social-cultural constraints, such as power and ideology, may also influence the way in which a classic is received and established in the target culture. In some cases, translated classics may generate changes, or even evolutions, in the receiving culture.

Third, this study addresses paratexts, a less discussed or not properly handled issue in previous studies on the translation of military classics. As a matter of fact, paratexts of great length are very common in almost all kinds of classics, in both source and target texts, and they often become an indispensable part of the core texts. Moreover, paratexts are places where translators, editors and other parties can more easily come in and exert their influence. Therefore, it is important to take both core texts and paratexts into consideration.

Last, this study adopts a mixed method of qualitative and quantitative analysis to investigate the translations of *AOW* as well as their reception in target cultures. Sample analysis will be conducted to come up with detailed analysis and spur in-depth discussion. At the same time, three sets of corpora for different purposes will be established. Statistics drawn from these self-built corpora will be used to identify the translation approaches, to map out the general reception trend and to provide a clear and tangible presentation of the social-cultural context.

The present study on the translation of *AOW* is necessary and significant in at least four aspects. To begin with, this study may enhance our understanding of the translation of strategic culture (especially military classics) and our awareness of the role translation plays in the global cross-cultural exchange. This book, with a systematic study on how and why Sun Tzu's military thinking is translated and received in the English-speaking countries, will add to the knowledge of how translation functions in military and social spheres. Next, the findings of this study will hopefully offer some insights to the different roles of the core text and its paratexts in cultural reconstruction in translation and prompt due attention to paratexts in classics. Furthermore, it is hoped that this study will provide some reference to find a feasible way to bridge the gap between investigations of translation approaches and examinations of the reception of translated texts in the target cultures. The established link between them may reveal more clearly the impact of translation approaches on a work's reception and may offer a new perspective from which we may judge translation approaches, such as domestication and foreignization, both synchronically and diachronically. Last but not least, the mixed method of qualitative and quantitative analysis will be suggestive in terms of methodology for future translation research dealing with similar issues.

References

Ames, Roger T. trans. 1993. *The Art of Warfare: Classics of Ancient China (the first English translation incorporating the recently discovered Yin-chueh-shan texts, with an introduction and commentary)*, 1st ed. New York: Ballantine Books.

Anonymous. 2015. "The Commencement of Research on *The Art of War* in Warring State Period 《孙子兵法》研究的发轫时期." Sun Zi Document Online Repository, 孙子兵法文献网. Accessed June 13, 2020. www.sunzi512.com/info_onenr.aspx?Page=1&classid=111&a=te.

Blanchot, Maurice. 1983. "Translating." *Sulfur (Translated from the French by Richard Sieburth)* 26: 82–86.

Booth, Ken. 1979. *Strategy and Ethnocentrism*. London: Croom Helm.

Booth, Ken. 1990. "The Concept of Strategic Culture Affirmed." In *Strategic Power: USA/USSR*, ed. Carl G. Jacobsen, 121–128. London: Palgrave Macmillan UK.

Calthrop, Everard F. trans. 1905. *Sonshi: The Chinese Military Classic*. Tokyo: Sanseido.

Chen, Bingfu, and M.W. Luke Chan. trans. 1998. *Sunzi on The Art of War and Its General Application to Business*. Shanghai: Fudan University Press.

Chohan, Chou-wing, and Abe Bellenteen. trans. 2003. *The Art of War: The Cornerstone of Chinese Strategy (Edited by Brant, Rosemary)*. Hod Hasharon: Astrolog Publishing House.

Cleary, Thomas. trans. 1998. *The Art of War Sun Tzu,* 1st edition. Boston and Shaftesbury: Shambhala.

Clements, Jonathan. trans. 2012. *The Art of War.* London: Constable.

Gagliardi, Gary. trans. 1999. *The Art of War: In Sun Tzu's Own Words.* Seattle: Clearbridge Publishing.

Giles, Lionel. trans. 1910. *On the Art of War: The Oldest Military Treatise in the World (Translated from the Chinese with Introduction and Critical Notes by Lionel Giles).* London: Luzac.

Giles, Lionel. trans. 2007. *Sun Tzu's The Art of War (Bilingual Edition Complete English and Chinese Text, Translated by Lionel Giles, with a New Foreword by John Minford).* Tokyo, Rutland, Vermont, and Singapore: Tuttle Publishing.

Gray, Colin S. 1999. *Modern Strategy.* Oxford: Oxford University Press.

Griffith, Samuel B. trans. 1963. *The Art of War (Preface by Liddell Hart).* New York: Oxford University Press.

Hagy, Jessica. 2015. *The Art of War Visualized: The Sun Tzu Classic in Charts and Graphs.* New York: Workman Publishing Company.

Howard, Russell D. 2013. *Strategic Culture.* MacDill Air Force Base: The JSOU (Joint Special Operations University) Press.

Huang, Haixiang黄海翔. 2009. "Orientalism and Retranslation of Chinese Classics: A Case Study based on the Descriptive Analysis of the Paratexts of John Minford's Version of *The Art of War* 东方主义对典籍复译的影响——基于副文本描述的《孙子兵法》Minford译本个案分析." *Journal of Yibin University* 宜宾学院学报 9 (5): 88–91. doi:10.3969/j.issn.1671-5365.2009.05.029.

Huang, Haixiang黄海翔. 2018. *Cultural Studies in English Translations of The Art of War* 《孙子兵法》英译的文化研究. Guangzhou 广州: Jinan University Press 暨南大学出版社.

Huang, J. H. trans. 2008. *The Art of War: The New Translation.* New York: Harper Perennial Modern Classics.

Huang, Liyun黄丽云. 2013. "On Comprehending and Translating Ancient Culture-loaded Words in *The Art of War* 《孙子兵法》中古代文化负载词的理解和翻译探析." *Journal of Leshan Normal University* 乐山师范学院学报 28 (3): 84–87.

Johnston, Alastair I. 1995. "Thinking about Strategic Culture." *International Security* 19 (4): 32–64.

Lau, D. C. 1965. "Some Notes on the 'Sun tzu' 孫子." *Bulletin of the School of Oriental and African Studies, University of London* 28 (2): 319–335.

Lefevere, André. 1992. *Translation, History, Culture: A Sourcebook.* London: Routledge.

Li, Jijun李际均. 1997. "*On Strategic Culture* 论战略文化." *China Military Science* 中国军事科学 (1): 165–167.

Li, Jijun李际均. 2005. "*The Art of War* and Strategic Culture 孙子兵法与战略文化." *Military History Research* 军事历史研究 (1): 165–167.

Li, Yuri李浴日. 1938. *A Comprehensive Study on Sun Tzu's Art of War* 孙子兵法之综合研究. Changsha 长沙: The Commercial Press 商务印书馆.

Lianeri, Alexandra, and Vanda Zajko. 2008. "Introduction: Still Being Read after so Many Years: Rethinking the Classic through Translation." In *Translation and the Classic: Identity as Change in the History of Culture*, eds. Alexandra Lianeri and Vanda Zajko 1–23. Oxford: Oxford University Press.

Lin, Wusun. trans. 2003. *The Art of War.* San Francisco: Long River Press.

Longhurst, Kerry A. 2004. *Germany and the Use of Force.* Issues in German Politics. Manchester: Manchester University Press.

Luo, Tian, and Meifang Zhang 罗天、张美芳. 2015. "Research on Translations of *The Art of War*: Past and Prospects《孙子兵法》翻译研究五十年: 回顾与展望." *Translation Quarterly* 翻译季刊 75: 50–65.

Luo, Ziye, trans.罗志野译. 1995. *100 Sun Tzu's The Art of War* 孙子兵法一百则. Taipei 台北: The Commercial Press at Taiwan 台湾商务印书馆.

Michaelson, Gerald A., and Steven Michaelson. trans. 2010. *Sun Tzu: The Art of War for Managers: 50 Strategic Rules,* 2nd ed. Avon, MA: Adams Media.

Minford, John. trans. 2002. *The Art of War (with an introduction and commentary).* New York: Viking.

Nylan, Michael. trans. 2020. *The Art of War: A New Translation.* 1st ed. New York: W. W. Norton & Company.

Richter, Gregory C. trans. 2004. *Sūn Zǐ Bīng Fǎ Sun Zi's Art of War (Pinyin Transcription, Gloss, and English Translation).* Kirksville, MO: Truman State University.

Sadler, Arthur L. trans. 1944. *Three Military Classics of China: The Art of War of Sun Tzu, The Precepts of War by Sima Rangju, Wu Zi on the Art of War.* Sydney: Australasian Medical Publishing Company, Ltd.

Sawyer, Ralph D. trans. 1993. *The Seven Military Classics of Ancient China (with a commentary).* Boulder: Westview Press.

Scobell, Andrew. 1999. "Soldiers, Statesmen, Strategic Culture and China's 1950 Intervention in Korea." *Journal of Contemporary China* 8 (22): 477–497. doi:10.1080/10670569908724358.

Scobell, Andrew. 2002. *China and Strategic Culture.* Carlisle: Strategic Studies University.

Shao, Qing邵青. 2013. "A Review on the Circulation of *The Art of War* Overseas 《孙子兵法》海外传播述评." *Military History Research* 军事历史研究 (4): 97–102.

Snyder, Jack L. 1977. *The Soviet Strategic Culture: Implications for Limited Nuclear Operations, RAND R-2154-AF.* Santa Monica, CA: The Rand Corporation.

Song, Zhongwei. 2012. "*The Art of War* in Retranslating Sun Tzu: Using Cultural Capital to Outmatch the Competition." *Translation and Interpreting Studies* 7 (2): 176–190.

Su, Guiliang, and Nosuke Azusen 苏桂亮、阿竹仙之助. 2009. *Different Editions of The Art of War in Japan* 日本孙子书知见录. Jinan 济南: Shandong Qilu Press 齐鲁书社.

Su, Guiliang苏桂亮. 2011. "Study on the English Versions of *The Art of War*《孙子兵法》英文译著版本考察." *Journal of Binzhou University* 滨州学院学报 27 (5): 149–156. doi:10.3969/j.issn.1673-2618.2011.05.027.

Tang, Zichang. trans. 1969. *Principles of Conflict: Recompilation and New English Translation with an Annotation on Sun Zi's Art of War.* San Rafael, CA: T. C. Press.

The Denma Translation Group. trans. 2001. *The Art of War: The Denma Translation.* Boston: Shambhala Publications.

Venuti, Lawrence. 2008. "Translation, Interpretation, Canon Formation." In eds. Alexandra Lianeri and Vanda Zajko. *Translation and the Classic: Identity as Change in the History of Culture* 27–51. Oxford: Oxford University Press.

Wang, Jinfeng王普丰. 2004. "*The Art of War*: Gem of Chinese Strategic Culture 《孙子兵法》是中国战略文化的瑰宝." *China Military Science* 中国军事科学 (6): 14–16.

Wang, Xiaoyin王晓莹. 2011. "A Study on the Translation of Culture-loaded Words in *The Art of War* from the Perspective of Power Discourse 权力话语下的《孙子兵法》文化负载词翻译研究." *Dongjing Wenxue* 东京文学 (5): 247–248.

Sui, Yun. trans. 2004. *Sunzi's Art of War: World's Most Famous Military Classic* (illustrated by Wang Xuanming). 10th ed. Singapore: Asiapac Books.

Wei, Qianqian魏倩倩. 2018. *Classic Translation and Dissemination: A Case Study of The Art of War* 典籍英译与传播——以《孙子兵法》为例. Beijing 北京: People's Press 人民出版社.

Wing, R. L. trans. 1988. *The Art of Strategy: A New Translation of Sun Tzu's Classic: The Art of War.* 1st ed. New York: Doubleday.

Wu, Rongzheng吴荣政. 1996. "The Impact of *The Art of War* on Europe and America 论《孙子兵法》对欧美的影响." *Journal of Xiangtang Teachers College (social science edition)* 湘潭师范学院学报(社会科学版) (4): 20–24.

Wu, Rongzheng吴荣政. 2004a. "Spread and Research of Master Sun's *Art of War* in Japan 《孙子兵法》在日本的传播与研究." *Guizhou Social Science* 贵州社会科学 (2): 90–95.

Wu, Rusong吴如嵩. 2004b. "To Promote the Research on *The Art of War* with a Focus on Strategic Culture." *China Military Science* 中国军事科学 (6): 22–24.

Wu, Sha吴莎. 2013. *Journey of Chinese Strategy to the West: A Study on the English Translations of The Art of War* 兵学西渐 《孙子兵法》英译研究. Beijing 北京: Tuanjie Press 团结出版社.

Xiao, Gang肖刚. trans. 2013. *The First Book for Leadership: The Art of War* 领导科学第一书 兵法 汉英对照. Beijing 北京: China Economy Press 中国经济出版社.

Yang, Yuying杨玉英. 2012. *A Study on the Translations of The Art of War in English-speaking World* 英语世界的《孙子兵法》英译研究. Chengdu 成都: Sichuan University Press 四川大学出版社.

Yang, Yuying杨玉英. 2017. *A Study on the Dissemination and Reception of The Art of War in English-speaking World.* 《孙子兵法》在英语世界的传播与接受研究. Beijing 北京: Xueyuan Press 学苑出版社.

Yu, Rubo, ed.于汝波主编. 2001. *A History of Research on The Art of War* 孙子兵法研究史. Beijing 北京: Military Science Press 军事科学出版社.

Yuan, Shibing. trans. 1987/1990. *Sun Tzu's Art of War: The Modern Chinese Interpretation (Annotated by Tao Hanzhang).* New York: Sterling Pub. Co.

Yuen, Derek M. C. 2014. *Deciphering Sun Tzu: How to Read The Art of War*: New York: Oxford University Press.

Zhang, Guojun章国军. 2014. *Sun Tzu's Journey to the West: Retranslation and Misreading of Classics* 《孙子》西行 名著复译与误读 中文、英文. Beijing 北京: Foreign Language Teaching and Research Press 外语教学与研究出版社.

Zhang, Linyu, and Caixia Li 张琳瑜、李彩霞. 2011. "A Comment on the English Version of *The Art of War* from the Perspective of Poly-system 多元系统理论视角评析《孙子兵法》术语英译." *Journal of Harbin University* 哈尔滨学院学报 (1): 111–114.

2   Relevant theories and concepts

A literature review

This chapter introduces some key theories and concepts that are vital to this study of the translation and reception of *The Art of War (AOW)* in the West, which include Critical Discourse Analysis, paratext and culture and reviews the previous relevant research, with a purpose to establish an analytical framework for the study of the translation and reception of *AOW*.

2.1 Critical Discourse Analysis

In this research, we view the translation of *AOW* as a discursive activity that shapes and is shaped by social and cultural practices. This viewpoint is informed by the theory of Critical Discourse Analysis, which is explained in this section.

2.1.1 An overview of Critical Discourse Analysis

Discourse is a rather important and complex term. According to a summary by Jaworski and Coupland (1999, 1–3), discourse can be used in three senses: the organization of connected text beyond the level of the sentence (e.g. Carter 1982, 184); language in use (e.g. Brown and Yule 1983); and a broader composite of linguistic and nonlinguistic social practices (e.g. Fairclough 1995; 2001). In this study, we follow the third usage and share the view that in modern societies, discourse "has taken on a major role in sociocultural reproduction and change" (Fairclough 1995, 2).

There are many ways to examine discourse, among which the Critical Discourse Analysis (thereinafter referred to as "CDA"), created by scholars such as Norman Fairclough, Van Dijk and Wodak, is among the most important approaches. In Fairclough's view, CDA is the study of "how texts work within sociocultural practice" (Fairclough 1995, 7), and "a resource for producing richer understanding and analysis of the relationship between discourse and other non-discursive facets of social processes and social change" (Fairclough 2003, 52). Van Dijk (2001, 352) defines CDA as the discourse analytical research "that primarily studies the way social-power abuse and

DOI: 10.4324/9781003025726-2

inequality are enacted, reproduced, legitimated, and resisted by text and talk in the social and political context".

Generally speaking, CDA focuses on relations between text, discourse and social factors, such as power relations, ideologies, institutions and social identities, as Fairclough and Wodak remark:

> CDA sees discourse—language use in speech and writing—as a form of "social practice". Describing discourse as social practice implies a dialectical relationship between a particular discursive event and the situation(s), institution(s) and social structure(s), which frame it: The discursive event is shaped by them, but it also shapes them. That is, discourse is socially constitutive as well as socially conditioned—it constitutes situations, objects of knowledge, and the social identities of and relationships between people and groups of people.
>
> (Fairclough and Wodak 1997, 258)

Fairclough and Wodak (1997, 271–280) summarize eight main tenets of CDA. Four of them are introduced briefly here as they are informative to our case study of the translations of *AOW*.

1 CDA addresses social problems. The basic claim of CDA is that major social, political and cultural processes and movements have a partly linguistic-discursive character. In other words, CDA analyzes the linguistic and semiotic aspects of social and cultural processes, structures, and problems.
2 Power relations are discursive. CDA "highlights the substantively linguistic and discursive nature of social relations of power in contemporary societies" (Fairclough and Wodak 1997, 272). When discussing discourse and power, Fairclough (2001, 36) stresses two major aspects of the relationship: power in discourse, and power behind discourse. Discourse can be viewed as "a place where relations of power are actually exercised and enacted" (Fairclough 2001, 36), and power can be found in a face-to-face spoken interview, or in cross-cultural discourse. To understand the power behind discourse, we have to explore how orders of discourse are themselves shaped and constituted by relations of social orders.
3 Discourse constitutes society and culture. An instance of language use may contribute somewhat to the process of reproducing and/or transforming society and culture. There are three broad domains of social life that may be discursively constituted: representations of the world, social relations between people, as well as people's social and personal identities (Fairclough 1992). It can be assumed that any part of any language text, spoken or written, simultaneously constitutes representations, relations and identities.
4 Discourse does ideological work. "Ideologies are particular ways of representing and constructing society which reproduce unequal relations of

power, relations of domination and exploitation" (Fairclough and Wodak 1997, 275). Ideology represents social reality and constructs identities that are linked to power, especially the collective identities of groups and communities. It is often the case that untrue or incorrect ideology underpins the constructions of society. Discourses are in most cases loaded with ideology, and they often help to spread certain ideology. To decide whether a particular discursive event carries ideological note, it is needed not only to analyze texts but also to consider how texts are interpreted and received and what social effects they have.

2.1.2 Three-dimensional analytical framework

Fairclough (1995, 25, 158; 2001, 21) has put forward a three-dimensional macro framework of CDA: (1) textual analysis; (2) discursive analysis, which includes the analysis of the processes of text production, distribution, and interpretation or consumption; and (3) socio-cultural analysis of the discursive event (see Figure 2.1).

The first dimension of CDA, namely textual analysis, involves the description of the formal properties of a text. A text is a product of discourse, while discourse is the whole process of social interaction of which a text is just a part (Fairclough 2001, 20). It is primarily a linguistic-cultural artifact and traditionally a piece of written or oral work, such as a poem, a speech, an interview, a novel or a book chapter. Textual analysis includes the examination of textual form, structure and organization at all levels: phonological, grammatical, lexical, and higher levels of textual organization, structures of argumentation and generic structures (Fairclough 1995, 7).

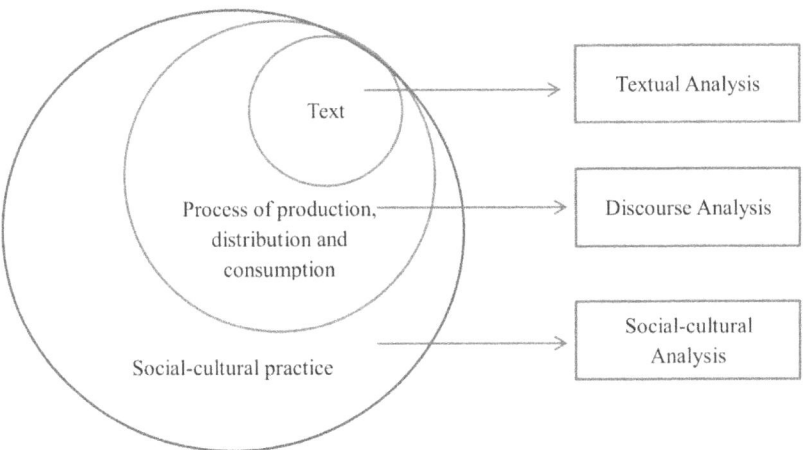

Figure 2.1 Three-dimensional framework of CDA (based on Fairclough 1995, 25; 2001, 21).

The second dimension of CDA is discourse analysis. This dimension has a crucial mediating role between the first and third dimension of CDA, between texts on one hand and society as well as culture on the other. According to Fairclough (1992), discourse practice involves the production, interpretation/ consumption and distribution of texts. Text production concerns issues such as how producers of text use resources and approaches, and how they draw upon and restructure orders of discourse to produce new configurations of genres and discourses. Text distribution or circulation can be investigated in terms of "chain" relationships within orders of discourse. There are more or less settled chains of discursive practices within and between orders of discourse across which texts are shifted and transformed in systematic ways.

Fairclough (2001, 134) points out that three questions need to be answered in discourse analysis: (1) the Context: what interpretation are participants giving to the situational and intertextual contexts? (2) Discourse Type: what discourse type(s) are being drawn upon; (3) Difference and Change: are answers to questions about context and discourse type different for different participants? And do they change during the course of the interaction?

Two aspects of context shall be covered here: situational and intertextual. Situational context is related to social orders and determines discourse types (Fairclough 2001, 119). Intertextual context is related to interactional history (Fairclough 2001, 119, 128–129). Discourses and the texts which occur within them have histories, they belong to historical series, and the interpretation of intertextual context is a matter of deciding which series a text belongs to, and what can be taken as a common ground for the texts.

There are four aspects to investigate discourse types: content, subject, relation, and connection. Content answers the questions of what's going on in the discourse: its activity, topic and purpose. Subject concerns the issue of who's involved, including the subject's positions and social identities in different situations. When it comes to the question of relation, the subjects positions mean something more dynamically, in terms of relationship to power, social distance and so forth. Connection answers the question of what's the role of language in the discourse. Language is used in an instrumental way as a part of a wider institutional and bureaucratic objective.

The third dimension of CDA is sociocultural analysis, also referred to as *explanation*. It aims to show how a discourse, as (part of) a social process, is "determined by social structures, and what reproductive effects discourses can cumulatively have on those structures, sustaining them or changing them" (Fairclough 2001, 134). Three issues can be addressed in this dimension: (1) Social determinants; what power relations at situational, institutional and societal levels help shape this discourse? (2) Ideologies; what ideologies are rooted in the discourse or what the discourse attempts to spread? (3) Effects; how is the discourse positioned in relation to struggles at the situational, institutional and societal levels? Does the discourse contribute to sustaining existing power relations, to challenging them or to transforming them?

A piece of discourse is embedded within sociocultural practice at a number of levels; in the immediate situation, in the wider institution or organization, and at a societal level (Fairclough 1995, 96). It is therefore assumed that any discourse will have determinants and effects at all three levels of situational, institutional and societal (Fairclough 1995, 134). In other words, a discourse is shaped by power relations, and contributes to societal struggles at these three levels.

In Fairclough's three-dimensional framework of CDA, text production is shaped by discourse practice and discourse practice is conditioned by socio-cultural practice. This model aims to "bring together linguistically-oriented discourse analysis and social and political thought relevant to discourse and language" and to serve "social scientific research", specifically "the study of social change" (Fairclough 1992, 62). CDA is extensively used by scholars in analysis of media discourse, political discourse, medical discourse, educational discourse etc.; however, Fairclough notes that there is still "a need to bring close textual analysis together with social analysis of organizational routines for producing and consuming texts" and distribution "is a relatively neglected issue which merits more attention" (Fairclough 1995, 9).

2.1.3 Critical Discourse Analysis and translation studies

Modern scholars in Translation Studies (hereinafter referred to as TS) have taken increasing interest in the complexities of power relations and ideo-logical impact in the production of translations. Some of them began to apply CDA into TS, bringing about insightful discussions of discourses, especially political and media discourses.

Schäffner (2004), for instance, makes a very compelling statement for interdisciplinary cooperation between translation studies (TS) and Political Discourse Analysis. She lists a number of examples from EU political dis-course where aspects such as lexical choice and conceptual metaphors in translations had contrasting effects in the British and German political and media environments. The translational choices illustrate that target language (TL) texts and discourses are framed by social and political structures and practices. With this argument, she establishes the important links between TS and CDA.

Valdeón (2005; 2007; 2008) is among the scholars who apply Fairclough's three-dimensional model to investigating ideological issues in translation. For instance, by drawing upon Fairclough's model of CDA, he analyzed the ideological implications of the lexical choices made in the news reporting on the 2004 Madrid bombings (Valdeón 2007). He compares the relevant texts from leading Spanish news websites, to nine English-speaking news websites and their respective Spanish versions. He discovered that the translation pro-cess had been shaped mainly by editorial notions of ideology, making the role of the text writers submissive to the political stance of the company, and changing a series of informative texts into controversial reports aimed at their target audience. The study provides an illustration of the importance of contrastive Critical Discourse Analysis in translation studies.

A research study by Kang (2007) takes a critical approach to news translation from the perspective of discourse analysis. By investigating an instance of how news stories about North Korea are translated by a South Korean news media company, Kang's paper aims to demonstrate that translation is itself the product of sociocultural, as well as textual, processes. Kang concludes that news translation is not a complete and accurate representation of the ST, but a recontextualization practice featuring institutional goals and procedures, together with tension and conflict between different representations, ideologies and voices.

Baumgarten (2010) incorporates CDA with corpus linguistics to compare 11 translations of Hitler's *Mein Kampf*. His analysis shows how the framing of different target texts reflects contrasting ideological positions on the way Hitler was to be represented. The corpus-assisted approach provides another meeting point for CDA and TS, which allows the investigation of ideology in translation to be balanced by a larger volume of data.

Zhang and Pan (2015) applied Fairclough's CDA model and systemic functional linguistics to the study of the translation of public notices about SARS in Macao. Their study explored how language is used by different governmental institutions in shaping their social power. With an investigation of speech roles, speech functions, modality types and modality orientation based on a comparison of the SARS notices and their translations, the authors argue that choices made in producing the texts reflect the governmental institutions' social roles and their relationship with each other and with the audience. Their work shows that different models of discourse analysis, such as SFL and CDA can be applied to translation studies in a supplementary way.

Al-Hejin (2012) proposes three methodological models for linking Fairclough's dialectical-relational approach to CDA with text-based approaches in TS. The first model emphasizes translation as re-writing by conducting a textual analysis of a translation and then analyzing the discursive and sociocultural practices of the translator operating in the target language domain. In this model, translators are both ideological agents and subject to orders of discourse that impact their work, and a translation can be viewed as a text in its own right which reflects the interests of a particular cultural group. The second model emphasizes translation as an intertextual chain by conducting a comparative Critical Discourse Analysis of the text's discursive and social practices of both the SL and TL domains. Analysts need to trace the motivation behind certain linguistic choices back to the source language and culture in order to gain a proper understanding of their intended function and discursive impact. The third model takes into consideration multiple versions to conduct a comparative analysis of two or more translations of one text (see Figure 2.2). These multiple translations could be different English translations of the same ST, or translations in different languages.

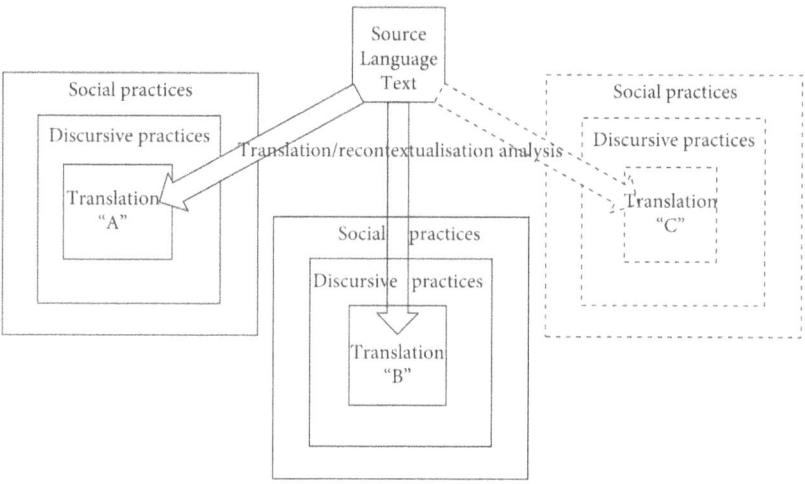

Figure 2.2 A comparative mode of CDA in translation studies (Al-Hejin 2012).

According to the above literature review, the CDA perspective can offer a critical understanding of the translation production. Based on the previous translation studies from the perspective of CDA, the present study tries to explore how *AOW* was translated, how the translation was received and what social cultural changes it generated in the target culture, and how *AOW* translation and reception are conditioned by the social-cultural context.

2.2 Text, paratext and translation

In the following section, Gérard Genette's paratext theory will be introduced, highlighting the importance of paratexts in CDA and TS. We believe that core texts and paratexts should both be given attention in the investigation of the translation of classic texts.

2.2.1 Concept of paratext

The concept of "paratext", coined by Gérard Genette (1997, 1), refers to the verbal or non-verbal materials accompanying a text, such as an introduction, notes, dedications, preface, foreword, figures, book jacket, pictures, epigraph, blurbs, etc.

According to Genette (1997, 1–15), in order to define the paratextual message, a number of features of the paratext should be taken into account: the paratext's spatial, temporal, substantial, pragmatic and functional characteristics. According to the location, paratexts can be classified as peritexts, which exist inside a book, or epitexts, which are outside a book. Most paratexts are verbal (or linguistic), some are iconic (or illustrations), some are material (everything which proceeds from the typographical choice) and others are factual (the fact of its existence makes commentary on the text and has bearing on its reception). In terms of temporal situation, paratexts can be anterior (appearing earlier than the text), original (appearing at the same time as the text), subsequent (appearing shortly after a text) or belated (appearing in a much later edition) paratexts. Due to the existence of different authors, we have authorial (by the author of the text), editorial (by editors or publishers) or allographic (by a third party) paratexts.

No matter how varied paratexts are, the relationship between a text and its paratext usually remain the same. Paratext is subordinate to the text in its status. Except for some rare cases, "the paratext, in all its forms, is a fundamentally heteronomous, auxiliary, discourse devoted to the service of something else which constitutes its right of existence, namely the text" (Genette 1997, 12). This relationship suggests that if we want to secure a better understanding of a text, we should not neglect its paratext.

2.2.2 Functions of paratext

In the CDA framework, texts are considered linguistic products of social events with causal effects (Fairclough 2003, 8). They can bring changes in our knowledge and belief and contribute to shaping social identities. Texts can also start wars and generate changes in education and industrial relations. Paratexts, though being subordinate in status, are extremely important to the social effects of a text, due to their purposes and functions.

In Genette's (1997, 1) theory, the purpose of paratexts is to ensure a text's presence in the world, its reception and consumption in the form of a book. Paratext is considered a threshold of interpretation, exerting a considerable influence on the reader's reception of the text (Genette 1997, 1–2). As a fringe of text, paratext constitutes "a zone between text and off-text, a zone not only of transition but also of transaction: a privileged place of a pragmatics and a strategy, of an influence on the public" (Genette 1997, 2). Macksey (1997, xvii–viii) summarizes Genette's theory, saying that paratext is about "the literary and printerly conventions that mediate between the world of publishing and the world of the text" and it is set in "its complex mediations between author, publisher, and audience".

The function of paratext is defined by the characteristics of its communicatory instance or situation, including the nature of the addresser, of the addressee, the degree of authority and illocutionary force of the message (Macksey 1997, 266). According to Genette (1997, 8–11), there are different types of illocutionary force for the paratext: informative (of pure information), interpretive (of the authorial or editorial intention, this is the cardinal function of most prefaces), directive (of advice or persuasion), commissive (of undertaking a promise) and performative (to accomplish what it describes). However, there is still another common illocutionary force that Genette didn't specify, namely, the expressive, which articulates the paratext author's attitude or emotions.

Paratexts, as supporting elements of texts, can also be included in the macro framework of CDA. This is because paratexts not only involves the process of interpretation, consumption and circulation of the core text, but also mediate the sphere of text and society with its illocutionary forces.

2.2.3 Paratext in translation studies

The study of paratexts is particularly important because they offer valuable insights into the presentation and reception of translated texts within the target historical and cultural climate (Kovala 1996, 120). As "each paratext addresses a culturally-specific moment and a culturally specific readership, it projects a singular version of the text through the lens of the chronotope (time/place) of its publication" (Watts 2000, 31). The importance and function of paratexts in translation therefore lies in at least four aspects: readers, translators, texts and cultures, with the fourth being the focus of our present study.

As Watts (2000, 31) points out, the first function of paratexts in all contexts is to help, attract and guide target readers. For instance, Zhang (2012a) discovered that the Chinese translators of *Ulysses* used paratexts (including more than 6,000 footnotes), together with the method of adaptation to render this linguistically and culturally "unreadable" book into a readable one for ordinary Chinese readers. The second function of paratexts is to display the translator's own role and intervention in the text (McRae 2012). The third function of paratexts is to help in the process of the presentation and reception of translated texts within the target historical and cultural climate (Kovala 1996, 120). The fourth function is one of cultural translations (Watts 2000, 31). Paratexts can serve as a tool to represent or reflect the prevailing target norms and signals an interface between competing ideological positions.

Many scholars have contributed to the investigation of paratexts' function of cultural translation. For instance, McRae (2012), in a content analysis of translators' prefaces to contemporary literary translations into English, discovered that the most predominant function of a preface is the promotion of understanding between cultures. Gerber (2012) has mapped out the shaping effect that paratextual material has on the representation of the Cultural Other in the target culture community, with corpus-based research about German translations of Australian children's fiction. Summers (2013) found that a paratext may reveal political stances and at the same time may be used to manipulate the reader. According to Pellatt (2013, 1–3), the cultural implications of paratexts, their cultural significance and political, ideological and commercial power have a complex and profound influence on translating and translated works. Appiah (2000, 427–428) proposes an alternative term for the use of paratexts, "thick translation", which "seeks with its annotations and its accompanying glosses to locate the text in a rich cultural and linguistic context", and which can be designed to perform an ideological function in the target culture, for instance, to show respect to the foreign culture and "to challenge directly the assumptions of western cultural superiority".

The issue of paratexts in translation has begun to gain increasing academic attention. However, most of the previous studies investigate paratexts in order to find out what cultural factors are contained in them and how culture makes impact on the text. Little research concerns how culture itself is translated or reconstructed in paratexts, and how paratexts serve as a site for cross-cultural exchange.

2.3 Culture and translation

Language, text and culture are always the central themes in translation studies. They are also the key elements in CDA. Texts carry messages about customs, knowledge and beliefs (Fairclough 2001, 93), and they can bring about changes in our knowledge, beliefs, attitudes, values and so forth (Fairclough 2003, 8). This means that texts not only embody cultural elements but also contribute to cultural changes.

2.3.1 Definition of culture

Culture is a rather complex term defined differently by schools of scholars with different research purposes from various perspectives. According to American anthropologists Kroeber and Kluckhohn (1952), there was already a list of 164 definitions of culture by the year 1952. One of the earliest definitions of culture was by Tylor (1871, 1) who proposed that: "Culture, or civilization, taken in its broad, ethnographic sense, is that complex whole which includes knowledge, belief, art, morals, law, custom, and any other capabilities and habits acquired by man as a member of society". Another definition by Matsumoto (1996, 16) is: "the set of attitudes, values, beliefs, and behaviors shared by a group of people, but different for each individual, communicated from one generation to the next".

In this study, we take the definition by Harris and Moran (1996): culture is "a set of knowledge, beliefs, values, religion, customs, acquired by a group of people and passed on from generation to generation". Culture is not only acquired by individual persons, but also sets its root in organizations, communities and societies. Furthermore, it can be passed on from one community to another with a different linguistic or cultural background.

Despite the differences in definitions of "culture", most scholars (e.g. Ferraro 1998; Hofstede, Hofstede and Minkov 2010, 6) are likely to agree on the following characteristics of culture: (1) Culture is socially learned, but not biologically inherited. It comes from an accumulation throughout the history of a given community and it manifests itself in forms of activity and behavior. (2) Culture is shared by members of a certain community. (3) Culture is an integrated system formed by various parts, which are usually interconnected, interrelated and interdependent. (4) Culture is not static but dynamic because it is "created and continually recreated by social relationships, processes, and negotiations involving actors from all parts of a society" (Avrami, Mason and La Torre 2000, 14). Since culture is an integrated system, if one component changes, it is likely that other parts will adjust. When a culture comes in contact with other cultures, interactions and changes occur. Therefore, culture is always in the state of flux, being not only a collection of things, but also a set of processes.

Scholars such as Mazrui (1996) and Martins and Martins (2001, 382) have pointed out that culture has different functions for individuals, organizations and societies. Three of the generally agreed functions of culture are: perception, identity and behavior patterns. Firstly, culture gives "meaning to social situations" and generates "social roles and normative behaviours" (Matsumoto 2007). The way we look at and comprehend the social and natural worlds is shaped by culture. Secondly, culture helps people to establish a sense of belonging to a certain group or community. It also determines how they see themselves and others (e.g., in terms of gender, age and ethnicity). And thirdly, culture provides a person a ready-made set of patterns or procedures for certain actions such as dining, dating, preaching, doing business and doing battles. It also makes it possible to anticipate others' actions.

2.3.2 Analytical levels of culture

As a hierarchy system, culture can be analyzed at different levels and studied both synchronically and diachronically. According to Hofstede, Hofstede and Minkov (2010, 6), culture can be classified into many levels, such as (1) a national level according to one's country; (2) a social class level associated with people's educational opportunities, occupation or profession; (3) an organizational level according to the way employees have been socialized by their work organization; (4) a generation level separating grandparents from parents from children; and (5) a gender level according to whether one was born as a girl or as a boy.

House (2012, 499) puts forward four analytical levels on which culture can be differentiated: the general human level, the societal level, the subgroup level, and the individual level. The first level differentiates human beings from animals. The societal level suggests that culture is the unifying, binding force that enables people to position themselves in systems of government, domains of activities, religious beliefs and values. The subgroup level is about various societal subgroups according to geographical region, social class and professional activity. At the individual level, it is the personal guidelines of thinking and acting, which is related to cultural consciousness, that enables a human being to be aware of what characterizes his or her own culture and makes it distinct from others.

Culture is not built in a day. It has to go through different stages before it becomes firmly established. According to Sperber (1996), culture can go through three different stages of representations: individual, public and cultural. At the very beginning, there exists a multitude of individual mental representations, most of which are ephemeral and privately conceived. These representations involve personal thinking, speaking or acting. They may include individual beliefs, intentions or preference. Only a subsection of these representations can come out publicly in the form of symbols, language or artefacts. They then become public representations, which are communicated to others in the social group and produce similar mental representations. This process features repeated communication to a wider audience. If a portion of public representations are communicated frequently enough to a considerable number of members within a particular social group, these representations may become firmly established and turn into cultural representations.

The above arguments about the analytical levels of culture have both merits and disadvantages. Hofstede, Hofstede and Minkov's (2010) classification is rather complicated and a bit confusing since it lacks a consistent criterion. The human level in House's analytical model is redundant since when we discuss culture, we are already referring to the human examples. Sperber's taxonomy of three stages of cultural representation is progressive and clear cut, however, the term "cultural representation" for the third stage is overlapping with the umbrella concept of "culture".

Taking into consideration the merits of the above-mentioned arguments, an analytical model of culture, mainly based on what is proposed by Sperber but using different names, can be established at three levels: individual, institutional (or organizational), and social. While Sperber (1996) stresses on the diachronic development, we believe that culture can be examined at these levels both from bottom-up and top-down, both synchronically and diachronically. It has to be noted that sometimes the division among these levels is relatively blurry.

2.3.3 Culture in translation studies

Culture is a key theme in translation studies. Generally speaking, there are two sorts of research on culture in translation. One school follows the "cultural turn" in the earlier 1990s from the publication of a collection of essays entitled T*ranslation, History, Culture* co-edited by Bassnett and Lefevere (1990). "Cultural turn" means the change of research paradigm in translation studies from the traditional linguistic approach to one that focuses on extra-textual factors related to social, cultural and historical contexts. It embraces the tools of cultural history and cultural studies.

The cultural approaches to translation extend the research scope from purely linguistic to the following: looking at the way cultural context influences translations and probing into the ideology and ethics that impact translations (Cunico and Munday 2007; Elliott and Boer 2012; Pérez 2003; Pym 2001); and examining the power negotiations translators are involved in the translation process (Álvarez and Vidal 1996; Tymoczko and Gentzler 2002).

According to Munday (2012), theories proposed by scholars using cultural approaches include: Rewriting Theory, which studies the power relations and ideologies existing in the patronage and poetics of literary and cultural systems that interface with literary translation (Lefevere 1992); Feminist Translation Theory, which approaches translation from a gender-studies angle (Santaemilia 2005; Simon 1996); and Postcolonial Translation Theory, which considers that translation has played an active role in the colonization process and in disseminating an ideologically motivated image of colonized peoples (Bassnett and Trivedi 1999; Niranjana 1992; Spivak 2004).

Other scholars investigate how culture itself is translated. To be specific, they try to examine how the culture embodied in ST is reconstructed in TT and how translations help to bring changes into the target culture. There are two perspectives, the macro and micro, in this school of research. The macro perspective considers how translation functions in the process of cross-cultural communication in the larger context of the text, for instance, in the target literary systems (Burke and Hsia 2007; Hung 2005; Peng and Rabut 2014; Schäffner and Kelly-Holmes 1995). It is generally acknowledged that "[t]ranslation wields enormous power in constructing representations of foreign cultures" (Venuti 1995b, 9). This perspective usually considers the translation of culture beyond the scope of texts. They study the intercultural communication occurring outside TT, but seldom go deeply into textual analysis.

In discussing the strategies for the translation of culture, the most frequently used the terms are domestication and foreignization, regarding the degree to which translators make a text conform to the target culture. The origin of these two terms can be traced back to Schleiermacher in 1813, who discusses two types of translation in which "[e]ither the translator leaves the author in peace, as much as possible, and moves the reader toward him. Or he leaves the reader in peace, as much as possible, and moves the author toward him" (Schleiermacher 1813/1992, 42). For Schleiermacher, the translator has to opt for one of these and keep to it consistently. Venuti (1995a) continues to revive and advance these two strategies in a modern sense. In Venuti's (1995a) view, domestication is the approach of making text closely conform to the target culture. With domestication, a fluent, and transparent style is adopted in order to minimize the strangeness of the translated text for the TL readers. Domestication may bring about greater acceptability, but it may involve the loss of information from the source text. Foreignization is the approach of retaining the foreignness, or the exotic flavor, of the source text, and it involves deliberately breaking the linguistic or textual conventions of the target language. Foreignization may mean the selection of a non-fluent, opaque style in order to provide TL readers with an "alien reading experience" (Venuti 1995a, 20).

Domestication is an inevitable practice in translation and it may result in the assimilation and then the loss of the foreign cultural essence to some extent, as Venuti (2008, 28) puts it: "[w]hen a classic is translated, furthermore, its very nature as a linguistic and literary artefact is fundamentally altered, along with the value it had acquired in the foreign culture where it was produced". To avoid such loss of meaning, Venuti (1995b) highly recommends that we rely on the approach of foreignization because it preserves cultural diversity and allows the readers access to foreign cultural values. Thus foreignizing can achieve what Venuti (1995b, 22–23) calls the "nonethnocentric translation", an ethically good translation that creates possibilities for cultural innovation, becomes "subversive of domestic ideologies and institutions", and reforms "cultural identities that occupy dominant positions in the domestic culture".

Some scholars, however, do not agree with Venuti's extreme preference for foreignization. Ellington (2003) believes that translators "fail if they go to either of these extremes" because extreme foreignization leads to a TT that is inaccurate or nonsensical, and extreme domestication may trivialize a ST and render its TT crude, anachronistic or silly. Mazi-Leskovar (2003) finds that in Slovenian translations of American prose great efforts are taken to keep the equilibrium between domesticating and foreignizing approaches, and such efforts involve both core text and paratexts. It is a wise and ethically sound choice to maintain the balance between domestication and foreignization.

Culture specific items (hereinafter referred to as CSIs), or culture-loaded terms, often become the topics in examining the translation of culture. There is a mass of case studies on the translation of CSIs. For instance, Aixela (1996) classifies the translation strategies of CSIs into conservation and substitution according to the degree of intercultural manipulation in translation. Wang (2012) conducts a comparative study of the translation of CSIs in the two English versions of *Hongloumeng* (*Dream of Red Mansions*) from the perspective of functionalist approaches. By classifying the translation approach and the corresponding methods of the two versions of the text Wang discovers that these two translations have different orientations for different purposes, with one mostly adopting foreignization, while the other more frequently employing domestication. Zhang's (2012b) study concerns the translation of site names in the Macau Historical Center. She finds that different translation strategies are applied in order to give tourists an access to the rich culture, unique history and dynamic nature of Macao.

The previous studies on the translation of culture are fruitful and feature a variety of theoretical frameworks, themes and methodologies. However, there are obvious limitations. Firstly, studies on the influence of culture upon translation far outnumber those on the translation of culture itself. Bassnett (1998, 138) has called for more investigation of the acculturation process that takes place between cultures, and more comparative study of the ways in which texts become cultural capital across cultural boundaries. Secondly, there is a lack of a model for the systematic research which includes both the translation of a source culture and its reception in a target culture. The diachronic reception of a translated culture in the target context is investigated even less. Thirdly, the previous studies on CSIs are featured with fragmentary and unsystematic sample analysis rather than a hierarchy and systematic framework of culture.

They routinely regard CSIs as "isolated occurrences in the text, usually at the word level" (Floros 2007). Sometimes a single example is used to represent the whole situation. For instance, Tu and Zhou (2008) hold that the translation of CSIs plays a positive role in cultural construction on the part of the translator. However, such a view looks a bit thin as it is supported by only one example concerning the translation of the word for the color red, "红", in Chinese.

The status quo of the research on translation of culture suggests a need to set up an integrated model for analysis on how a source culture is reconstructed within the TT, how it is consumed and received in the TT and what are the social and cultural factors behind its translation and reception.

2.4 Summary

In this chapter, we have firstly reviewed the theory of CDA, paying special attention to the three-dimensional framework proposed by Fairclough (1995; 2001). CDA constitutes an effective means through which text, discourse and social-cultural context can be systematically investigated. Our review has revealed that CDA has also been extensively applied in TS.

Secondly, there was a brief introduction of the concept of paratext as well as its functions. In TS, paratext has already become a theme for many researchers. However, the interactional relationship between text and paratexts, and the role of paratexts in the translation of culture shall be given more attention in the following chapters.

Last but not least, in both CDA and TS, culture is a key issue. On one hand, texts represent culture and their production is influenced by cultural factors; on the other hand, texts may eventually lead to changes in target culture. Therefore, it is possible and necessary to investigate, with an integrated model of CDA, the reconstruction of a source culture embodied in a classic within the TT, its reception in and influence upon the target culture, and the social-cultural factors behind such translation and reception.

References

Aixela, Javier F. 1996. "Culture-Specific Items in Translation." In *Translation, Power, Subversion*. Vol. 8 of *Topics in Translation*, eds. Román Álvarez and M. C.-Á. Vidal, 52–78. Clevedon, Bristol and Adelaide: Multilingual Matters.

Al-Hejin, Bandar. 2012. "Linking Critical Discourse Analysis with Translation Studies: An Example from BBC News." *Journal of Language and Politics* 11 (3): 311–335.

Álvarez, Román, and M. C.-Á. Vidal, eds. 1996. *Translation, Power, Subversion*. Vol. 8 of *Topics in Translation*. Clevedon: Multilingual Matters.

Appiah, Kwame A. 2000. "Thick Translation." In *The Translation Studies Reader*, ed. Lawrence Venuti, 417–429. New York: Routledge.

Avrami, Erica C., Randall Mason, and Marta de La Torre. 2000. *The Values and Heritage Conservation: Research Report*. Los Angeles: Getty Conservation Institute.

Bassnett, Susan. 1998. "The Translation Turn in Cultural Studies." In *Constructing Cultures: Essays on Literary Translation*. Vol. 11 of *Topics in Translation*, eds. Susan Bassnett and André Lefevere, 123–140. Clevedon: Multilingual Matters.

Bassnett, Susan, and André Lefevere. 1990. *Translation, History and Culture*. London & New York: Pinter.

Bassnett, Susan, and Harish Trivedi, eds. 1999. *Post-colonial Translation: Theory and Practice. Translation Studies*. London: Routledge.

Baumgarten, Stefan. 2010. *Translation as an Ideological Interface: English Translations of Hitler's "Mein Kampf"*. Saarbrücken: Verlag Dr. Müller.

Brown, Gillian, and George Yule. 1983. *Discourse Analysis. Cambridge Textbooks in Linguistics*. Cambridge: Cambridge University Press.

Burke, Peter, and R. P. Hsia. 2007. *Cultural Translation in Early Modern Europe*. Cambridge: Cambridge University Press.

Carter, Ronald. 1982. *Linguistics and the Teacher*. London: Routledge & Kegan Paul.

Cunico, Sonia, and Jeremy Munday. 2007. *Translation and Ideology: Encounters and Clashes*. London and New York: Routledge.

Ellington, John. 2003. "Schleiermacher Was Wrong: The False Dilemma of Foreignization and Domestication." *The Bible Translator* 54 (3): 301–317.

Elliott, Scott S., and Roland Boer. 2012. *Ideology, Culture, and Translation. Society of Biblical Literature Semeia Studies* 69. Atlanta: Society of Biblical Literature.

Fairclough, Norman. 1992. *Discourse and Social Change*. Cambridge: Polity.

Fairclough, Norman. 1995. *Critical Discourse Analysis: The Critical Study of Language. Language in Social Life Series*. London: Longman.

Fairclough, Norman. 2001. *Language and Power*. 2nd ed. *Language in Social Life Series*. Harlow, Essex: Longman.

Fairclough, Norman. 2003. *Analysing Discourse: Textual Analysis for Social Research*. London: Routledge.

Fairclough, Norman, and Ruth Wodak. 1997. "Critical Discourse Analysis." In *Discourse as Social Interaction*. Volume 2 of *Discourse Studies: A Multidisciplinary Introduction*, ed. Teun A. van Dijk, 258–284. London: SAGE.

Ferraro, Gary P. 1998. *The Cultural Dimension of International Business*. 3rd ed. New Jersey: Prentice Hall.

Floros, Goregios. 2007. "Cultural Constellations and Translation." MuTra 2007—LSP Translation Scenarios: Conference Proceedings.

Genette, Gérard. 1997. *Paratexts: Thresholds of Interpretation* (translated by Jane E. Lewin; foreword by Richard Macksey). Vol. 20 of *Literature, Culture, Theory*. Cambridge: Cambridge University Press.

Gerber, Leah. 2012. "Marking the Text: Paratexual Features in German Translations of Australian Children's Fiction." In *Translation Peripheries: Paratextual Elements in Translation*, eds. Anna Gil Bardají, Pilar Orero and Sara Rovira Esteva, 43–62. Bern: Peter Lang.

Harris, Philip R., and Robert T. Moran. 1996. *Managing Cultural Differences*. Houston: Gulf Publishing Company.

Hofstede, Geert H., Gert J. Hofstede, and Michael Minkov. 2010. *Cultures and Organizations: Software of the Mind: Intercultural Cooperation and Its Importance for Survival*. 3rd ed. New York and London: McGraw-Hill.

House, Juliane. 2012. "Translation, Interpreting and Intercultural Communication." In *The Routledge Handbook of Language and Intercultural Communication. Routledge Handbooks in Applied Linguistics*, ed. Jane Jackson, 495–519. London: Routledge.

Hung, Eva. 2005. *Translation and Cultural Change: Studies in History, Norms, and Image Projection.* Vol. 61 of *Benjamins Translation Library*. Philadelphia, PA: John Benjamins Publishing.

Jaworski, Adam, and Nikolas Coupland. 1999. *The Discourse Reader*. London: Routledge.

Kang, Ji-Hae. 2007. "Recontextualization of News Discourse." *The Translator* 13 (2): 219–242.

Kovala, Urpo. 1996. "Translations, Paratextual Mediation, and Ideological Closure." *Target* 8 (1): 119–147.

Kroeber, Alfred L., and Clyde Kluckhohn. 1952. *Culture: A Critical Review of Concepts and Definitions.* Vol. 47 of *Papers of the Peabody Museum of American Archaeology and Ethnology*. Cambridge, MA.: Harvard University.

Lefevere, André. 1992. *Translation, Rewriting, and the Manipulation of Literary Fame. Translation Studies*. London: Routledge.

Macksey, Richard. 1997. "Foreword" In *Paratexts: Thresholds of Interpretation,* ed. Gérard Genette. trans Jane E. Irewin, xi–xxii Cambridge: Cambridge University Press.

Martins, Nico, and E. Martins. 2001. "Organisational Culture." In *Organisational Behaviour: Global and Southern African Perspectives*, eds. Stephen P. Robbins, Timothy A. Judge, Aletta Odendaal and Gert Roodt, 381–384. Pinelands, Cape Town: Pearson Education South Africa.

Matsumoto, David. 1996. *Culture and Psychology*. Pacific Grove, CA: Brooks/Cole Publishing Company.

Matsumoto, David. 2007. *Culture, Emotion, and Expression*. New York: New York Academy of Sciences.

Mazi-Leskovar, Darja. 2003. "Domestication and Foreignization in Translating American Prose for Slovenian Children." *Meta* 48 (1–2): 250–265.

Mazrui, Al. 1996. "Perspective: The Muse of Modernity and the Quest for Development." In *The Muse of Modernity; Essays on Culture as Development in Africa*, eds. Philip G. Altbach and Salah M. Hassan, 1–18. Trenton, NJ: Rica World Press.

McRae, Ellen. 2012. "The Role of Translators' Prefaces to Contemporary Literary Translation into English: An Empirical Study." In *Translation Peripheries: Paratextual Elements in Translation*, eds. Anna Gil Bardají, Pilar Orero and Sara Rovira Esteva, 68–82. Bern: Peter Lang.

Munday, Jeremy. 2012. *Introducing Translation Studies: Theories and Applications*. 3rd ed. Oxford & New York: Routledge.

Niranjana, Tejaswini. 1992. *Siting Translation: History, Post-structuralism, and the Colonial Context*. Berkeley, CA and Oxford: University of California Press.

Pellatt, Valerie, ed. 2013. *Text, Extratext, Metatext and Paratext in Translation*. Newcastle upon Tyne: Cambridge Scholars Publishing.

Peng, Xiaoyan, and Isabelle Rabut, eds. 2014. *Modern China and the West: Translation and Cultural Mediation*. Vol. 2 of *East Asian Comparative Literature and Culture*. Leiden: Brill.

Pérez, María C. 2003. *Apropos of Ideology: Translation Studies on Ideology, Ideologies in Translation Studies*. Manchester, UK, and Northampton, MA: St. Jerome Publishing.

Pym, Anthony, ed. 2001. *The Return to Ethics*. vol. 7, no. 2: Special issue of *The Translator*. Manchester: St. Jerome Publishing.

Santaemilia, José, ed. 2005. *Gender, Sex and Translation: The Manipulation of Identities*. Manchester, UK, and Northampton MA: St. Jerome Publishing.

Schäffner, Christina. 2004. "Political Discourse Analysis from the Point of View of Translation Studies." *Journal of Language and Politics* 3 (1): 117–150.

Schäffner, Christina, and Helen Kelly-Holmes, eds. 1995. *Cultural Functions of Translation*. Clevedon: Multilingual Matters.

Schleiermacher. 1813/1992. "On the Different Methods of Translating." In *Theories of Translation: An Anthology of Essays from Dryden to Derrida*, eds. Rainer Schulte and John Biguenet, 36–54. Chicago and London: University of Chicago Press.

Simon, Sherry. 1996. *Gender in Translation: Cultural Identity and the Politics of Transmission. Translation Studies*. London: Routledge.

Sperber, Dan. 1996. *Explaining Culture: A Naturalistic Approach*. Oxford: Blackwell.

Spivak, Gayatri C. 2004. "The Politics of Translation." In *The Translation Studies Reader*. 2nd ed., ed. Lawrence Venuti, 397–416. New York and London: Routledge.

Summers, Caroline. 2013. "What Remains: The Institutional Reframing of Authorship in Translated Peritexts." In *Text, Extratext, Metatext and Paratext in Translation*, ed. Valerie Pellatt, 9–32. Newcastle upon Tyne: Cambridge Scholars Publishing.

Tu, Guoyuan, and Hui Zhou. 2008. "Culture-specific Item Translation and the Translator's Cultural Awareness: A Case Study of the Two English Versions of the Translation of 'Red' in *Hongloumeng*." *Journal of Central South University (Social Science)* 14 (6): 891–894.

Tylor, Edward B. 1871. *Primitive Culture: Researches into the Development of Mythology, Philosophy, Religion, Art, and Custom*. London: John Murray.

Tymoczko, Maria, and Edwin Gentzler. 2002. *Translation and Power*. Amherst: University of Massachusetts Press.

Valdeón, Roberto A. 2005. "The 'Translated' Spanish Service of the BBC." *Across Languages and Cultures* 6 (2): 195–220.

Valdeón, Roberto A. 2007. "Ideological Independence or Negative Mediation: BBC Mundo and CNN en Español's (translated) Reporting of Madrid's Terrorist Attacks." In *Translating and Interpreting Conflict*, ed. Myriam Salama-Carr, 97–118. Brill: Rodopi.

Valdeón, Roberto A. 2008. "Anomalous News Translation: Selective Appropriation of Themes and Texts in the Internet." *Babel* 54 (4): 299–326.

van Dijk, Teun A. 2001. "Critical Discourse Analysis." In *The Handbook of Discourse Analysis. Blackwell Handbooks in Linguistics*, eds. Deborah Schiffrin, Deborah Tannen and Heidi E. Hamilton, 352–371. Malden, MA and Oxford: Blackwell Publishers.

Venuti, Lawrence. 1995a. *The Translator's Invisibility: A History of Translation. Translation Studies*. London and New York: Routledge.

Venuti, Lawrence. 1995b. "Translation and the Formation of Cultural Identities." In *Cultural Functions of Translation*, eds. Christina Schäffner and Helen Kelly-Holmes, 9–25. Clevedon: Multilingual Matters.

Venuti, Lawrence. 2008. "Translation, Interpretation, Canon Formation." In *Translation and the Classic: Identity as Change in the History of Culture. Classical Presences*, eds. Alexandra Lianeri and Vanda Zajko, 27–51. Oxford: Oxford University Press.

Wang, Yuefang. 2012. "Exploring Cultural Transmission and Translation Strategies in the Perspective of Functionalist Approaches: A Case Study of the Two English Versions of Hongloumeng." *Babel* 58 (4): 471–487.

Watts, Richard. 2000. "Translating Culture: Reading the Paratexts to Aimé Césaire's Cahier d'un retour au pays natal." *TTR: Traduction, Terminologie, Rédaction* 13 (2): 29–45.

Zhang, Meifang. 2012a. "Annotation and Adaptation: A Case Study of a Chinese Translation of Joyce's *Ulysses*." *Translation Quarterly* 64: 32–54.

Zhang, Meifang. 2012b. "Reading Different Cultures Through Cultural Translation: On Translation of Site Names in Macau Historic Centre." *Babel* 58 (2): 205–219.

Zhang, Meifang, and Hanting Pan. 2015. "Institutional Power in and Behind Discourse: A Case Study of SARS Notices and Their Translations Used in Macao." *Target* 27 (3): 387–405.

3 Research methodology and data

One framework, two methods and three corpora

This chapter introduces the methodology and data of the study. It firstly presents an overall analytical framework for the study based on the relevant theories and concepts reviewed in Chapter 2. Specific analytical procedures of this study are then outlined. Finally, it describes data source, corpus building and the tools used for corpus analysis in this study.

3.1 Overall analytical framework and procedure

In order to investigate how Chinese strategic culture is reconstructed in the English translations of *The Art of War*, how it is received in the target culture and how to identify the factors that influence its translation and reception, an analytical framework is proposed, as shown in Figure 3.1. This analytical framework, drawing upon Fairclough's three dimensions of CDA, includes three major parts of investigation: textual analysis, discourse analysis and social-cultural analysis.

In this framework, both the ST and TTs, published in the form of books, are considered as textual products of certain social events. The ST is believed to be a military teaching presented by Sun Tzu to a king he was serving. It has been passed on through generations, annotated by a large number of scholars and esteemed as a strategic canon. As a result, it embodies the core values of Chinese strategic culture. Textual analysis aims to identify how Chinese strategic culture is represented in the ST and TTs through linguistic means such as military terms, figures of speech and syntactical structures.

By means of discourse analysis, this study will examine the process of how the TTs are received, including how they are interpreted, quoted and circulated in the target culture. We need to find out to what extent the translations of *AOW* have been incorporated into Western military and non-military discourse. Furthermore, we also need to systematically identify the way in which the TTs are received and what influences the translations brought into the Western discourse.

Social-cultural analysis explores factors behind *AOW*'s translation and reception. It discusses the translator–reader relationship, the ideological impact and power relations. This process aims to investigate why *AOW* is translated and received in the target culture as it is.

DOI: 10.4324/9781003025726-3

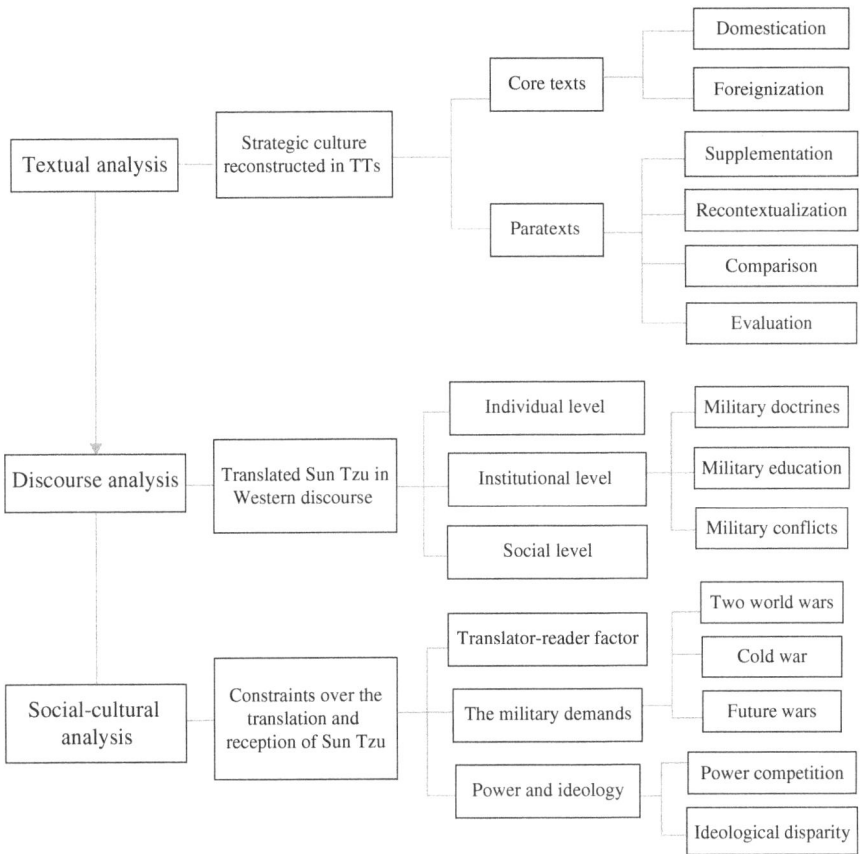

Figure 3.1 An overall analytical framework.

As there are multiple versions of STs and TTs of *AOW*, it is impossible and unnecessary to include all of them in one single piece of research. Therefore, one typical version of ST and two well-known TTs were included in the study. This may constitute a basis for a comparison between ST and TTs, and another comparison between the two TTs. In this sense, this study also follows the comparative mode of CDA in translation studies suggested by Al-Hejin (2012) (see Figure 2.2).

3.1.1 Textual analysis: culture reconstructed in core texts and paratexts

In the stage of textual analysis, we will investigate both the core text and paratexts of the ST and TTs to find out how ancient Chinese strategic culture is reconstructed.

In the core text, specific attention is given first of all to the translation approaches of culture-specific items (CSIs). CSI is a linguistic unit (usually a word or phrase) that represents a small portion of a certain culture individually. A collection of CSIs will map out the general content of a certain culture. Therefore, how CSIs are translated may become a key issue in translating a text into a target culture. In this study, approaches adopted for translating CSIs can be examined to showcase how the strategic culture represented in the core texts is rendered into TL counterparts. Based on Venuti's (1995) classification of translation approaches for culture, namely foreignization and domestication, and Wang's (2012) categorization of translation methods mentioned in Section 2.3, a framework for the analysis of CSIs was established, as shown in Table 3.1. It has to be noted that foreignization and domestication are usually termed as two translation "strategies", however, in this book, they are referred to as two translation "approaches" so as to avoid confusion with the term "strategy" in the military sense that is most frequently used.

To obtain a clear understanding of the strategic culture reconstruction in the paratexts of TTs, the cultural content of the ST and TTs is analyzed in a comparative mode. Based on the illocutionary forces of paratexts proposed by Genette (1997, 8–11), and considering the fact that this study focuses on strategic culture moving from one social-cultural context into another, four approaches with which paratexts may deal with a foreign culture are proposed: supplementation, recontextualization, comparison and evaluation.

Supplementation means to add some extra information into the paratexts to make up for a deficiency, and/or to clarify the obscurity concerning cultural

Table 3.1 Approaches to translate CSIs

Approach	*Method*	*Definition*	*Examples*
Foreignization	Calque	To translate literally, word by word	轻车: light chariots
	Transliteration	to translate proper names according to pronunciation	赵将李牧: The Chao general Li Mu
Domestication	Omission	No translation nor interpretation given	
	Paraphrasing	To explain the meaning with a culture-free word	曲制: organization
	Generalization	To translate the meaning with a more general word	荀卿: a philosopher
	Substitution	To replace SL CSI with a cultural counterpart in TL	石: bushel 镒: hundredweight

references in the core text. Frequently, many cultural elements are neither included nor clearly articulated in the SL text because they are common background knowledge shared by ST readers and there is limited space particularly in the core text. However, such an omission of cultural references may cause difficulty for readers from a different linguistic and cultural background. Supplementation is therefore helpful to compensate for such a loss in the translated core text. This approach closely resembles what Genette (1997, 8–11) called the informative function of paratexts since both suggest the addition of new information.

Recontextualization suggests interpreting some ancient cultural elements in a modern setting, and/or flavouring those from the foreign land with local ingredients, and/or explaining local specialities in global, common terms. In a ST, contextualization is frequently seen in paratexts as "each paratext addresses a culturally-specific moment and a culturally-specific readership, it projects a singular version of the text through the lens of the chronotope (time/place) of its publication" (Watts 2000, 31). Therefore, it is most likely that each time a ST text is translated anew, the paratexts need to be contextualized in a new manner so as to cater to the needs of the target readers and bring the text closer to them. Recontextualization is similar to what (Baker 2006, 112) referred to as temporal and spatial reframing in narratives.

Comparison involves comparing the source and target culture, which is a commonly used approach to reconstruct a culture in the paratexts of a TT. In translation, the Cultural Other is always reconstructed in relation to the Cultural Self, either in an explicit or an implicit manner. In some cases, elements of the source culture are compared explicitly in paratexts with its target counterparts in order to identify the commonality or difference. The revealed similarities and differences may help readers to measure their relations and better comprehend the source culture. The comparative approach can offer a basis for further judgement or evaluation on the source cultural elements, which is crucial to the reception of the source culture in the target culture.

Evaluation refers to the expression of the stance, attitude and viewpoints of the translator or other authors of paratexts. Evaluation occurs more frequently in paratexts than in core texts since translators are often asked to be neutral and impartial in their translations of core texts. However, in prefaces and introductions, translators or other parties tend to and are usually allowed the freedom to express their attitudes toward the translated culture or cultural elements, either acclamation or criticism, satisfaction or hatred. Such evaluations, either positive or negative, are more likely to exert a great influence upon readers and become crucial for the reception of the culture reconstructed in the core texts.

These four approaches for culture reconstruction in paratexts constitute a comprehensive and useful toolkit for us to do textual analysis while we are observing how the Chinese strategic culture is reconstructed in paratexts of *AOW* translations.

3.1.2 Discourse analysis: reception of the translated Sun Tzu

In discourse analysis, the reception of the translated Chinese strategic culture will be examined at three levels: individual, institutional and social. These levels are based on the analytical level of culture, as mentioned in Section 2.3.

The first step is to investigate how the translated strategic culture was read, interpreted, cited or applied by individual persons. It is reasonably believed that individuals will make different contributions to the interpretation and reception of a translated text. Some of the privately conceived interpretations of a translation are ephemeral. Others may last longer and become well-known. If the translated strategic culture has been read and circulated by a sufficient number of individual military officers and strategic thinkers and has become relatively influential among a relatively larger number of institutional members for a long time, it reaches the second level, namely the institutional level, as discussed in the following paragraph.

The second step aims to establish how the translated *AOW* was received at the institutional level: among military groups, organizations and institutions. At this stage, the interpretation and circulation of the translation is a group behaviour practiced for a common purpose. Therefore, the investigation will cover a wide range of military institutions, particularly the four branches of the US military: Army, Navy, Air Force and Marine Corps.

In the third step, reception of the translated Chinese strategic culture will be examined at the social level. It attempts to find out whether the translated *Art of War* has gone beyond the military institutions to reach a far wider range of social communities. If translations of *AOW* are extensively received in non-military society, it is proven true that *AOW* has been re-canonized in the target culture, with a more extensive influence.

To ensure the comprehensiveness of the data and integrity of this study at this level, great efforts are invested to gather as much relevant data as possible. Since our focus is on the translation and reception of strategic culture, most of the data was found in relevant Western military texts. A wide range of text types was covered. For instance, journal papers, monographs, book chapters, military doctrines written or compiled by military academic and educational institutions were included in the corpora set up for the present study. For the investigation of *AOW* reception in non-military fields, texts from business, management, sports, entertainment, et cetera were collected for examination.

Intertextuality can be used as an important parameter to measure the reception of the translated *AOW* in the target culture. Intertextuality, in this scenario, can provide concrete evidence that translations of Sun Tzu have been interpreted, quoted and applied in the Western discourse. To be specific, if we can identify, in the Western military texts, as many quotations from the translated *AOW* as possible, either direct or indirect, we can find out how translations of Sun Tzu have received in Western discourse. More detailed factors to be considered include who is quoting the translated Sun Tzu as well as what is quoted, how frequently and under what circumstances and to what effect the quotes are used. With all these quotes, the links between the translation and reception can be established.

3.1.3 Social-cultural analysis: constraints over the translation and reception

Social-cultural analysis, the third stage of this study, will investigate the social-cultural factors that may impact the translation and reception of *AOW*. The analysis is conducted at three levels: situational, institutional and societal.

At the immediate situational level, translator–reader factors will be examined, including the social identities of the translators and readers, as well as their relationships, to understand why *AOW* is translated in such a way as described in the textual analysis.

At the institutional level, the needs of military organizations are investigated to explore why *AOW* is received and circulated in such a way as described in the discourse analysis. Special attention is given to the demands of strategic innovation among military institutions, which include educational organizations (particularly universities), research institutes and combat forces.

Strategic thinking is closely related to power relations and ideology. At the societal level, it is essential to look into the hidden constraints behind the translation and reception of *AOW*. According to CDA, power and ideology are two prevailing factors that govern almost every social event. In this study, we will try to identify which power relations and prevailing ideologies are playing leading roles in propelling the translation and reception of *AOW*.

3.2 Research methods, data collection and processing

This study incorporates two research methods, namely qualitative and quantitative analysis, with an effort to be as valid and credible as possible.

3.2.1 Qualitative and quantitative analysis

For qualitative analysis, case analysis and example analysis are employed. At the stage of textual analysis, two translations were used in the case analysis to showcase the translation of strategic culture approaches in core texts and reconstruction methods in paratexts. These two translations were carefully chosen because they are the most quoted and popular translations among more than 60 English versions, according to our statistics as shown in Section 6.1. A detailed and in-depth analysis of these most representative translations will reveal their successful approaches in dealing with strategic culture. At the same time, ample examples are drawn from the ST, TTs and other relevant texts so as to provide answers to the research questions listed in Section 1.4.

The quantitative analysis in this study depends largely on the self-built corpora designed according to our analytical framework. In other words, this study is essentially corpus-assisted, with statistics gathered from the corpora and used in the three stages—textual analysis, discourse analysis and social-cultural analysis. Quantitative statistics concerning translation methods, the quotation frequencies of a particular translated sentence and words or phrases are drawn from self-built corpora.

To answer the needs of this study, three corpora were designed and set up. The first corpus was designed for the textual analysis, specifically the investigation of the translation methods of the ancient Chinese strategic culture, particularly Sun Tzu's strategic thinking. The second corpus was used for the investigation of the reception of *AOW* translation in the Western military discourse as well as for the relevant influential social-cultural context. The third corpus was built for the exploration of the reception of *AOW* translations in the Western non-military discourse.

3.2.2 *Corpus 1: ST, TT1 and TT2*

The first corpus consists of a source text and two translations, designed for the textual analysis. Among many editions, *The Art of War Commented by Ten Scholars* (孙子十家注), edited by Sun Hsing-yen (孙星衍, 1752–1818), a famous scholar of the Qing Dynasty was chosen. Sun Hsing-yen's edition came out in 1877, and became one of the most authoritative and frequently reprinted Chinese versions. Consisting of the core text by Sun Tzu in 13 chapters and six paratexts as well as comments by 11 commentators, this representative edition became the basis of many translations. In this study, Sun Hsing-yen's edition, reprinted in 1937 by Beijing Commercial Press (Sun 1937), is used as the source text (hereinafter referred to as ST) for this study.

As mentioned in Chapter 1，at least 60 English versions of *AOW* appeared in a span of 100 years，each by a different translator, and these translations were also republished in numerous translation editions. On one hand, it is impossible and unnecessary to include all the translations in one study. On the other hand, it is reasonable and necessary to map out the diachronic process of translation, as well as to form a basis for comparison. Therefore, the two most popular and representative English translations were selected: Lionel Giles' (1910) as the first target text (hereafter referred to as TT1) and Samuel Griffith's (1963) as the second target text (hereafter referred to as TT2). Both translations clearly state that they are based on ST, the Chinese version edited by Sun Hsing-yen, which forms a solid foundation for comparative analysis.

TT1, entitled *Sun Tzu on The Art of War: The Oldest Military Treatise in the World*, was translated by Lionel Giles (1875–1958) and is considered the first scholarly translation and a particularly successful example of cross-cultural communication. Since its debut in London in 1910, it has been reprinted numerous times in various editions. From 2006 to 2010 alone, about 30 editions of Giles' translation were published, and other 11 books adapted from it (Su 2011, 150). Many successive English versions by other translators were also influenced by this groundbreaking version to a greater or lesser extent.

TT2, *Sun Tzu: The Art of War*, was translated by Samuel B. Griffith (1906–1983) and was first published in 1963 by Oxford University Press. It has been

Table 3.2 Corpus 1: ST, TT1 and TT2

Text	Core text	Paratexts	Total
ST (in Chinese characters)	6,107 (4.59%)	127,046 (95.41%)	133,153 (100%)
TT1(in English words)	10,472 (13.71%)	65,922 (86.29%)	76,394 (100%)
TT2(in English words)	9,554 (14.74%)	55,271 (85.26%)	64,825 (100%)

chosen for inquiry because it is widely considered as the most representative English version of Sun Tzu's text and is listed in the Chinese translation series of the UNESCO Collection of Representative Works. TT2 also boasts various editions and astounding sales figures.

All the core texts and paratexts of ST, TT1 and TT2 were scanned and transformed into digital files. The length of the ST is measured by the number of Chinese characters and the lengths of TT1 and TT2 are measured by the number of English words. As a result, a corpus for research was built up, as shown in Table 3.2.

The SL core text, which expounds on military theory, has only about 6,100 Chinese characters. In contrast, paratexts of the ST are voluminous, with a length more than 20 times that of the core text. TT1 contains a core text of about 10,400 English words translated from the SL core text, and multiple paratexts in 65,922 words, more than six times the length of the translated core text. TT2 includes the translation of the SL core text in about 9,000 English words, and paratexts in more than 64,800 words, about seven times the length of the translated core text.

In order to investigate the translation approaches and methods in the core texts, a bilingual comparative corpus was built. The core texts of ST, TT1 and TT2 were aligned manually, sentence by sentence, and input into CUC ParaConc 0.3, a free software designed for bilingual and multilingual corpus retrieval. The CUC Paraconc V0.3 enables the searched items to be displayed simultaneously in the source text and two or more translated versions, allowing for a comparative analysis of ST, TT1 and TT2.

A sample search with the CUC Paraconc V0.3 is shown in Figure 3.2. To understand how the Chinese military term "奇正" is translated into TT1 and TT2, the SL key word "奇正" was fed in and as a result four items came out in three sets of sentences numbered 1, 2 and 3. In sentence set 1, there is a SL sentence "三军之众，可使必受敌而无败，奇正是也" which contains the key word "奇正", while the other two are the translated sentences which contain the translations of "奇正" respectively: "manoeuvres direct and indirect" in TT1 and "the extraordinary and the normal forces" in TT2.

3.2.3 Corpus 2: Sun Tzu's reception in military discourse

To map out the general trend of reception of the translated Sun Tzu in the Western military discourse, firstly we searched and identified the military texts

Figure 3.2 A sample search with the CUC Paraconc V0.3.

in which *AOW* translations are quoted and used. The focus of this study is on the written English texts produced in the West, especially in English-speaking countries. The time span for our search begins with the year 1905, when the first English translation was published, and ends in the year 2020. The scope of the search covered a wide range of texts: journal papers, monographs, edited volumes, military doctrines, theses, news articles, book reviews, research reports and so on. Hard copies and e-files from libraries and on-line documents were gathered. While searching online resources from ProQuest, Taylor & Francis Online, Google and other websites, key words such as "Sun Tzu", "*The Art of War*", "strategy" and "military" were used in various combinations.

The preliminary search established the fact that Sun Tzu is mentioned in a very large number of English texts, and it is impossible and impractical to examine all of them. Therefore, a criterion for selecting texts was set up as follows: (1) texts that address strategy, war and other military topics were selected; (2) texts with titles including "Sun Tzu" or "*The Art of War*" were highly preferred; (3) texts directly and extensively quoting Sun Tzu from identifiable translation sources were prioritized; (4) texts indirectly quoting Sun Tzu with a clear indication of translations were also included. Those texts referring to Sun Tzu but lacking references to translations are not included in the corpus.

These relevant texts were then sorted out, classified and transformed into digital files for use. As a result, a corpus of Sun Tzu's reception in Western military discourse was created that consists of 410 texts in about 8,595,868 words, including 190 journal papers, 58 monographs, 47 military doctrines, 39 book chapters from edited volumes, 27 research reports, 25 theses and 24 news articles and online articles (see Table 3.3).

Table 3.3 Corpus 2: Sun Tzu's reception in Western military discourse (1910–2020)

Type	Number of texts	Number of words	Period	Main sources
Journal papers	190	1,097,236	1910–2020	*Joint Force Quarterly; Comparative Strategy; Military Review; Parameters; Small Wars Journal; Strategic Analysis*
Monographs	58	4,239,856	1954–2015	Strategic Studies Institute, Army War College; Routledge; Frank Cass; Greenwood Publishing Group
Military doctrines	47	1,744,184	1982–2015	Joint Chiefs of Staff; Headquarters US Marine Corps, Department of The Navy; Headquarters US Air Force Doctrine Center; Headquarters, Department of the Army, Joint Forces Command
Book chapters	39	341,847	1940–2015	Strategic Studies Institute, Army War College; Pergamon Press Inc.; Stanford University Press
Research reports	27	389,690	1986–2013	Strategic Studies Institute, Army War College (US); Naval War College (US); US Army Command and General Staff College, Air War College; National War College; Army War College
Theses	25	753,587	1978–2019	United States Army War College (US); Naval Postgraduate School (US); Air University (US),
News articles & Online articles	24	29464	1966–2020	United Service Institute Journal; USA Today, Los Angeles Times; Washington Post, The Times, New Statesman
Total	410	8,595,868	1910–2020	

The texts selected for the corpus are most likely representative of the translated Sun Tzu, since the overwhelming majority of them are from reliable sources: prestigious publishers, influential journals, strategic research institutes and military universities.

Most of the selected journal papers are from a great variety of professional and prestigious journals in strategic issues. Some papers are from journals issued by the US Marine Corps, Army, Air Force and Navy, while others were published in the UK, Australia and Canada. Among the 190 journal papers are 29 from the *Marine Corps Gazette*, 16 from *Joint Force Quarterly*, 11 from *Parameters*, 10 from *Comparative Strategy*, 12 from *Military Review*; 4 from *Jane's Defense Weekly* and 4 from *Small Wars Journal. Marine Corps Gazette* is a professional journal for US Marines that was founded in 1916 at the Marine Corps Association by then-Colonel John A. Lejeune. Its mission is to provide a forum for the exchange of ideas that advances knowledge, interest and esprit in the Marine Corps, and a vehicle for the dissemination of military art and science among Marines. *Joint Force Quarterly* is another important military periodical and is issued by the US National Defense University Press in Washington D.C. in support of the secretary of defense and the chairman of the joint chiefs of staff. It is the chairman's joint military and security studies journal designed to inform and educate national security professionals on joint and integrated operations, national security policy and strategy and developments in joint military education to help America's military and security apparatus meet challenges. *Comparative Strategy* is an international journal co-sponsored by the US National Institute for Public Policy and the Center for Strategic Studies, University of Reading. *Military Review* is one of the premier US military magazines, providing a forum for original thought and debate on the art and science of land warfare and other issues of current interest to the US Army and the Department of Defense.

Most of the 58 monographs were published in the US and UK. "Monograph" here refers to a sole authored book with an ISBN number. The Strategic Studies Institute, Army War College (USA) contributed 13 monographs to our corpus; Routledge, 7, and Frank Cass 4. The Strategic Studies Institute (SSI) was established in Carlisle, Pennsylvania, by the US Army War College to conduct geostrategic and national security research and analysis in support of the college and its curricula, the leaders of the Army and the Department of Defense, and the wider strategic community. It publishes three to five books per year, including edited compilations. SSI also publishes "Carlisle Papers in Security Strategy" by Army War College students. In addition, SSI provides an opportunity for government civilians, guest scholars and military officers to explore a wide range of strategic military issues by issuing "SSI Monographs".

Among the 47 military doctrines in the corpus, 15 are from the US Joint Chiefs of Staff, 10 from Headquarters, Department of the Army, and 8 from the HQ Air Force Doctrine Center. Other doctrines included were issued by forces in the UK, Canada and Australia.

The book chapters in the corpus are from edited volumes composed of papers by different authors. Nearly all the book chapters are from prominent publishers, such as Oxford University Press, Stanford University Press, Pergamon Press and Princeton University Press. In addition, nine of these chapters are from SSI of the US Army War College.

Research reports are written by staff members or authored by the students in strategy research institutes or military colleges. They are neither books published with an ISBN number, nor dissertations written for the purpose of a master's or doctoral degree. The corpus includes four reports from SSI of Army War College and four from Naval War College.

After Corpus 2 of these 410 military texts had been properly prepared, a quantitative analysis was conducted with the help of the software Antconc 3.4.4 (Anthony 2014) to present an overview of Sun Tzu's reception in the West. Issues investigated included: the diachronical development of the reception of *AOW*, whether it was received positively or negatively, how frequently Sun Tzu is quoted, which translation is best received, which strategic principles are most frequently quoted and how the core text and paratexts are quoted. Specifically, I looked into the number of quotes from Giles' and Griffith's translation in order to find out whether their translation approaches and methods in addressing the core texts and paratexts were effective (see Figure 3.3 for an example search for the quotation of "all warfare is based on deception" from *AOW* translations).

With an investigation of these quotations, Corpus 2 was used as the main data source to discuss how the translated Sun Tzu has been received in the

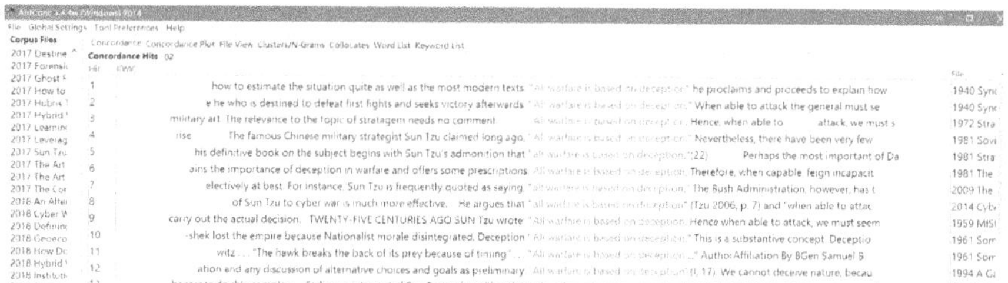

Figure 3.3 A sample search for quotations with Antconc 3.43.

Western military discourse and why. I looked into who—and which institutions or social groups quoted—discussed, applied or innovated the translated Sun Tzu. My purpose here was to map out the possible stages the translated Sun Tzu has undergone in the Western strategic culture.

Finally, through Corpus 2, some key terms closely related to "war", "country", and "ideology" were searched so as to further investigate in what circumstances Sun Tzu was received and quoted. My purpose here was to discover the constraints on the translation and reception of Sun Tzu in the West, such as institutional needs, power struggles and ideological disparities. I also selected some important strategists or military texts for sample analysis.

3.2.4 Corpus 3: Sun Tzu's reception in non-military discourse

In addition to Corpus 2, Corpus 3 was built up to investigate how *AOW* has been received in spheres other than the military (see Table 3.4). Texts that elaborate the issue of strategy in its general sense, in business, marketing, management and sports et cetera were also sorted out and included in Corpus 3. This corpus consists of 81 texts with more than 1 million English words.

Table 3.4 Corpus 3: Sun Tzu's reception in non-military discourse (1990–2020)

Disciplines	Number of texts	Number of words	Period	Main sources
Business and management	55	901,602	1990–2019	Mc-Graw Hill, Palgrave Macmillan, Penguin Group, Adams Media Corporation, Industrial Management & Data Systems, Oxford Leadership Journal, Journal of Strategic Management, Journal of International Business Ethics, Multinational Business Review
Sports and entertainment	12	128, 130	1992–2011	Tuttle Publishing, Coach and Athletic Director, The Daily Mirror, Los Angeles Times, The Daily Telegraph
Other disciplines	14	317,296	1995–2020	Adams Media Corporation, New Horizons in Education, strategic Impact, American Journal of Public Health, Journal of American Physicians and Surgeons, CRC Press
Total	81	1,218,898	1990–2020	

A similar analytical process was performed with the help of corpus search software such as Antconc 3.2.4.

With the help of several software programs, quantitative analysis can be conducted to find out the overall number of quotes; how Giles' and Griffith's English versions, respectively, have been quoted; and the most frequent citations of the translated Sun Tzu. Detailed example analysis can also be done to compensate for the deficiencies in corpus-based analysis.

References

Al-Hejin, Bandar. 2012. "Linking Critical Discourse Analysis with Translation Studies: An Example from BBC News." *Journal of Language and Politics* 11 (3): 311–335.

Anthony, Laurence. 2014. *AntConc*. Tokyo: Waseda University.

Baker, Mona. 2006. *Translation and Conflict: A Narrative Account*. London and New York: Routledge.

Genette, Gérard. 1997. *Paratexts: Thresholds of Interpretation (translated by Jane E. Lewin; foreword by Richard Macksey)*. Vol. 20 of *Literature, Culture, Theory*. Cambridge: Cambridge University Press.

Giles, Lionel. trans. 1910. *Sun Tzu on the Art of War: The Oldest Military Treatise in the World (Translated from the Chinese with Introduction and Critical Notes by Lionel Giles)*. London: Luzac.

Griffith, Samuel B. trans. 1963. *The Art of War (translated by Samuel B. Griffith, preface by Liddell Hart)*. New York: Oxford University Press.

Su, Guiliang苏桂亮. 2011. "Study on the English Versions of *The Art of War*《孙子兵法》英文译著版本考察." *Journal of Binzhou University* 滨州学院学报 27 (5): 149–156.

Sun, Hsing-yen, ed. 1937. *The Art of War Commented by Ten Scholars*. Beijing 北京: Commercial Press 商务印书馆.

Venuti, Lawrence. 1995. *The Translator's Invisibility: A History of Translation*. *Translation Studies*. London and New York: Routledge.

Wang, Yuefang. 2012. "Exploring Cultural Transmission and Translation Strategies in the Perspective of Functionalist Approaches: A Case Study of the Two English Versions of *Hongloumeng*." *Babel* 58 (4): 471–487.

Watts, Richard. 2000. "Translating Culture: Reading the Paratexts to Aimé Césaire's Cahier d'un retour au pays natal." *TTR: Traduction, Terminologie, Rédaction* 13 (2): 29–45.

4   Strategic culture reconstruction in core texts and paratexts of English translations

This chapter consists of four sections. The first section examines the Chinese strategic culture represented in the source text, with an introduction to the challenges from the source text that translators have to cope with: ancient and concise yet polysemous language, military terms and principles embodying Chinese military insights and philosophical underpinnings, and multiple voluminous paratexts, The second section investigates, in Giles' English version, how Chinese strategic culture is translated in the core text and how it is reconstructed in the paratexts. The third section, similarly, discovers the approaches Griffith has taken to reconstruct Chinese strategic culture in the core text and its paratexts. The fourth section is a short summary of the findings from the investigation.

To render a military classic such as *The Art of War* into another language, translators need to cross linguistic borders, to deal with paratexts attached to the core text and transcend cultural barriers. Section 4.1 introduces the challenges that translators must cope with.

4.1 Challenges in translating *The Art of War*: language, strategy and paratexts

The intrinsic quality of *AOW* lies in Sun Tzu's profound military terms, fundamental principles for strategy and unique style of representation. In this section, we introduce the challenges in translating the strategic culture represented in the SL core text and its paratexts. The analysis includes three aspects: linguistic features, military terms and principles, and multiple paratexts.

4.1.1 Linguistic features

The original version of *AOW* by Sun Tzu is written in an ancient, polysemous, concise and yet vivid Chinese language, and even Chinese readers may find it a rather demanding task to acquire a clear and thorough understanding, not to mention translators.

DOI: 10.4324/9781003025726-4

AOW was written in ancient Chinese about 2,500 years ago. Although a large number of ancient words in the ST share the same spelling as their modern counterparts, they often convey different meanings and, in some cases, make up parts of speech. For instance, "全国" in modern Chinese means "the whole country", but in *AOW* it means "take the enemy's country whole and intact"; "全军" in modern Chinese means "the nation" or "national", while in *AOW* it means "to capture an army entire".

Conciseness is a key linguistic feature of *AOW,* since texts in ancient China were usually written by brush on bamboo slips or on silk, which necessitated brevity. Unlike a modern Chinese word, which is often composed of two or more Chinese characters, one-character words are very typical of ancient Chinese. This is true of ST. According to our statistics drawn from Corpus 1, among the 50 high-frequency words which occur 16 times or more, 40 are one-character words, which makes up an overwhelming majority (90%); while only 10 are two-character words (c.f. Appendix 2, a sample list of high-frequency Chinese words in *AOW*).

Conciseness can be easily spotted in several aspects. At the lexical level, Sun Tzu's military thoughts are represented in highly pithy words that may harbor rich implications and need several modern phrases or sentences to explain. At the syntactical level, a large number of sentences omit their subjects, objects or connectives. Such omissions make it more difficult to figure out the meanings intended by Sun Tzu. Furthermore, Sun Tzu often omits the reasoning steps in his argument, and comes directly to his conclusion. He prefers affirmative and powerful aphorism-like statements over step-by-step substantiation or concrete examples of battles and generals in history for his argument. As Handel noticed:

> *The Art of War* reads more like a manual intended as a compact guide for the "prince" or higher-ranking military commander... Sun Tzu, for the most part presents the reader with his conclusions... some of Sun Tzu's concepts are more implicit (or are arrived at intuitively)... for the reader of *The Art of War*, acceptance of conclusions is the principal requirement.
>
> (Handel 2001, 16–17)

One specific example is Sun Tzu's statement about the foreknowledge in strategy. Sun Tzu firmly announces that "故曰：知己知彼，百战不殆；不知彼而知己，一胜一负；不知彼不知己，每战必殆" (Hence the saying: If you know the enemy and know yourself, you need not fear the result of a hundred battles. If you know yourself but not the enemy, for every victory gained you will also suffer a defeat. If you know neither the enemy nor yourself, you will succumb in every battle). This axiom is declared in a parallel structure, and highlights the importance of knowing both enemy and self. However, Sun Tzu does not follow the reasoning procedure to deduce his statement from a major premise nor to generalize it from a series of famous battles. Instead, Sun Tzu just proposes his self-evident maxim in strategic theory. Such succinct maxims are abundant in the core text.

Polysemy is another important linguistic feature of *AOW*. According to our statistics drawn from Corpus 1, among the 180 one-character content words with high frequency, 74 are polysemous, usually with different parts of speech, which constitutes a big portion of the total, 41%. These polysemous words require special attention from translators. For instance, the word "上" in "故明君贤将，能以上智为间者，必成大功" is an adjective, meaning "first-class", "excellent", while "上" in "善攻者，动于九天之上" is a noun, meaning "the place above". Another word, "计", in "将听吾计，用之必胜", is a noun referring to "stratagem"; while "计" in "料敌制胜，计险易、远近，上将之道也" is a verb meaning "calculate" or "count". Another typical example is 兵; with a frequency of 70 in the SL core text, it has multiple meanings in different collocations or contexts, ranging from "war" to "weapons", "soldiers", "army" and "tactics". "兵" in "兵众" means "soldiers", in "钝兵挫锐", "weapons", in "屈人之兵", "army", in "兵久而国利者", "warfare" and in "上兵伐谋", "stratagem". Such multiplicity in meaning constitutes undoubtedly a great obstacle for any ambitious translator.

Vividness is another language feature in the core text. Although fond of terse language, Sun Tzu was never a dull writer. To illustrate a large number of abstract military terms, *AOW* resorts to poetic and figurative language. Vividness permeates the text, as Sun Tzu expounds his military theory with a large number of rhetorical devices. In total, there are approximately 273 cases of rhetorical figures among 431 sentences, which suggests about 6 rhetorical figures in every 10 sentences (as shown in Appendix 3). These rhetorical devices constitute a distinctive feature of *AOW* and add to the persuasive power of the text.

These figures fall into about 16 types: antithesis, parallelism, metaphor, simile, anaphora, hyperbole, repetition, antimetabole, contrast, allusion, anti-climax, comparison, metonymy, anadiplosis, paradox and synecdoche. Among them, antithesis takes the largest share, with 97 cases amounting to 36% of the total. Parallelism takes second place with 48 cases of parallel structures used at the word, phrase and sentence level, amounting to 18%. Both antithesis and parallelism contribute to the rhythmic movement of the text and increase its persuasive power.

In addition to antithesis and parallelism, metaphors and similes saturated with Chinese flavor are frequently seen. Our investigation found that there are 28 metaphors and 19 similes and in the core text (as shown in Appendix 4). For instance, Sun Tzu compares the onset of troops to the rush of a torrent and an officer's timely decision to the swoop of a falcon which strikes its victim. He also compares the quickness of an army's movement to the rapidity of a running hare. In Sun Tzu's phrases, an army shall be as "swift as the wind", as "calmly majestic as the forest", as "plundering like fire", and as "steady as the mountains". These four similes were appreciated by sixteenth century Japanese general Takeda Harunobu, who had them embroidered on his battle banners as slogans (Griffith 1963, 173). Vivid similes and metaphors Sun Tzu's strategic thinking is easy to comprehend and long-remembered by his readers. However deeply rooted in the Chinese mode of thinking and culture, they are alien and fresh to English readers and consequently become a delicate issue in translation.

The above analysis shows that *AOW* is written in ancient, concise and highly rhetorical language, with a large number of polysemous words. On one hand, its conciseness and vividness guarantee *AOW* the quality of a military classic and invites the perusal and interpretation by scholars and generals of later generations, who continue to furnish it with paratexts to reinforce its status part of the military canon. On the other hand, such linguistic style is a great challenge for any translator who wishes to produce an accurate and elegant translation.

4.1.2 Military terms and principles

As a masterpiece on strategy, the core text of *AOW*, with 13 chapters, displays "the concentrated essence of wisdom on the conduct of war" (Hart 1963, vi). Each chapter deliberates on one topic: namely, estimates, waging war, offensive strategy, dispositions, strategic advantage, weakness and strength, maneuvers, tactics, marches, terrain, grounds, attack by fire and the use of spies (for details, refer to Section 1.1). These chapters develop an organic and systematic treatise to guide rulers and military officers in the intelligent prosecution of war, which is expressed in military terms and can be condensed to several principles.

Military terms

Terms are essential in establishing a theory and presenting an argument. The text of *AOW* is no exception. CSIs in strategy and military terms abound in Sun Tzu's 13 chapters. Therefore, a thorough investigation of the military terms used is necessary and significant in providing a bird's-eye view of strategic culture in *AOW.*

Based on Fu's (2001) classification, military terms in *AOW* can be sorted and summarized into the following five categories: strategy and tactics, organization, supply and armament, terrain, and battles and officers (see Table 4.1). There are about 195 strategic terms, of which 119 (61% of the total) are about strategy and tactics, 28 (14%) are about military supply and armament, 24 (13%) are about military terrains, 16 (8%) are about military organization and 8 (4%) are about famous battles and military figures in ancient Chinese military history.

These military terms that are essential to Sun Tzu's military thinking exist in a hierarchical system and most of them are interlinked. Some terms are superordinate, each of which may include several hyponyms, such as 全胜 (complete victory), 处军(positioning the army), 五危 (five dangerous faults)

Table 4.1 Military terms in ST

Category	Military terms	Amount
Strategy and tactics	爱民、败之道、北、崩、必生、必死、兵、兵势、兵形、不仁、不虞之道、称、驰、处军、处斥泽之军、处平陆之军、处山之军、处水上之军、道、度、夺气、夺心、饵兵、伐兵、伐交、伐谋、法、反间、分数、忿速、刚柔、攻、攻城、归师、诡道、行军、火队、火积、火库、火人、火辎、节、九变、九天、举军、距闉、军形、军争、开阖、利害、廉洁、量、乱、乱军、縻军、庙算、内间、破国、破军、破旅、破卒、破伍、奇正、强弱、穷寇、全国、全军、全旅、全卒、全伍、全胜、仁、仁义、锐卒、上将之道、生间、胜、势、守、数、死间、四军之利、天、天地、天之灾、围师、委军、五行、五间、五危、先知、陷、乡导、形、形名、虚、实、佯北、夷关、因间、阴、阳、阴阳、勇怯、用兵、迂、迂直之计、争地、政之道、知己、知彼、直、治变、治力、治乱、治气、治心、走、火攻	119 (61%)
Supply and armament	蔽橹、兵、车甲之奉、车、大车、带甲、盾、輠辐、革车、箕秆、戟、甲、胶漆之材、金鼓、旌旗、粮食、橹、弩、其、千金、轻车、丘牛、矢、委积、符、钟、胄、主用、辎重	28 (14%)
Terrain	九地、挂形、重地、隘形、百里、交地、绝地、绝涧、圮地、轻地、衢地、散地、死地、天井、天牢、天罗、天隙、天陷、通形、围地、险形、远形、支形、地形	24 (13%)
Organization	霸王、官道、曲制、天下、诸侯、军、旅、卒、伍、士卒、吏、大吏、三将军、三军、主、司命	16 (8%)
Battles and officers	刿、黄帝、吕牙、四帝、吴人、伊挚、越人、诸	8 (4%)

and 火攻 (attack by fire). The term 全胜 covers victory at five grades: 全国 (take the enemy's country whole and intact), 全军 (capture an army entire), 全旅 (capture a battalion entire), 全卒 (capture a company entire), and 全伍 (capture a squad entire). 处军 includes four situations: 处山之军 (positioning the army in mountains), 处水上之军 (positioning the army by water), 处斥泽之军 (positioning the army in salt marshes) and 处平陆之军 (positioning the army on plains). 五危 refers to five faults that a general can exhibit that may lead to a defeat: 必死 (recklessness, which leads to destruction), 必生 (cowardice, which leads to capture), 忿速 (a hasty temper, which can be pro- voked by insults), 廉洁 (a delicacy of honour, which is sensitive to shame) and 爱民 (over-solicitude for his men, which exposes him to worry and trouble). 火攻 is one of the typical tactics in ancient Chinese warfare, though not neces- sarily popular in modern times. According to Sun Tzu, there are five ways of attacking with fire: 火人 (to burn soldiers in their camp), 火积 (to burn stores), 火辎 (to burn baggage-trains), 火库 (to burn arsenals and magazines) and 火队 (to burn the lines of transportation).

In addition to the hierarchical order of these terms, there is also a close connection among them. For instance, a close association exists between 先知(foreknowledge) and 间 (spies). Sun Tzu attaches great importance to knowing both the self and the enemy prior to the actual fighting. One way of knowing the enemy, he thinks, is to use spies. In the 13th chapter, Sun Tzu puts forward a classification of 五间 (five kinds of spies): "故用间有五：有因间，有内间，有反间，有死间，有生间" (Hence the use of spies, of whom there are five classes: Local spies; inward spies; converted spies; doomed spies; surviving spies). The rest of the chapter defines these five kinds of spies, discusses the advantages of each, narrates the circumstances under which these spies are used and lists successful examples of spies. It is on the basis of these terms for spies that Sun Tzu's strategic thinking about espionage is established and represented in the text.

It must be noted that some military terms are well-defined and amply explained, while others are furnished with no definition. Since most of these terms are unique in Chinese culture, they pose a great challenge for translators.

Strategic principles

Based on the ancient military terms, Sun Tzu proposes several fundamental principles of strategy, and these principles constitute the key elements of ancient Chinese strategic culture (c.f. Guo 1984, 7–19; Li 1991, 1–6; Zhou 1992, 5–15). Particularly, these principles can be distilled into five vital ones that best summarize Sun Tzu's strategic wisdom. These five key principles are most frequently valued and cited in China.

Principle 1: Victory without fighting. In Sun Tzu's view, military force is the last resort after other efforts fail. The best policy is to counterstrike the enemy's strategy; the next best is to break up his coalitions; a good one is to fight the enemy's army in the field; and the worst is to attack the enemy's cities. When fighting has to happen, the ultimate and ideal goal is to subdue the enemy without battles rather than to annihilate the enemy's army or destroy his cities. In gaining victory, it is best to keep the enemy's army intact. Sun Tzu thinks that the one who wins by using minimum force and preserving his own strength is the state's treasure. It is evident from these statements that *AOW* is strongly marked with a fundamental distaste for warfare although it is a military treatise. Essentially, Sun Tzu is against militarism.

Principle 2: A holistic approach to strategy. Sun Tzu developed a holistic approach to warfare, emphasizing statecraft by combining military strategy with politics, diplomacy, psychology and economy. There is no country that can benefit from protracted warfare, because if the campaign is prolonged, the resources of the nation will be drained. Sun Tzu also drew attention to the importance of moral influences, leadership and psychological issues in war.

Principle 3: Foreknowledge and information. Sun Tzu stresses, before the fighting begins, the importance of adequate knowledge of both the enemy and the self for in-depth analysis, considerate planning and the formulation of an overall strategy. Such knowledge and planning involve various aspects, such as terrain, weather, training status and morale. Sun Tzu also focuses on the importance of developing ample military information sources through espionage and specifies the different types of spies and how to efficiently make use of them.

Principle 4: Dialectical view on warfare. Sun Tzu takes a dialectical view, with a series of paired concepts, to analyze the enemy and battlefield situations. These mutually defined and interrelated concepts are offense and defense, void and actuality, strong and weak, advance-retreat, *ch'i* (indirect) and *cheng* (direct), *ying* and *yang*, cold and warm. These primitive dialectical concepts, most likely from the Taoist worldview of the unity of opposites, provide the dynamic guidelines for generals to deliberate the ever-changing battle situations (Sawyer 1993, 130–131; Tao 2007, 188–189). For instance, one of the most important paired military concepts proposed by Sun Tzu is *ch'i* (奇) and *cheng* (正). In essence, *cheng* tactics suggest employing troops in the direct, normal or expected ways, such as massive frontal assaults, while *ch'i* tactics are realized by means of flexible forces in indirect, extraordinary and unconventional ways, such as flanking and rear attacks. There is also a reciprocal relation between *ch'i* and *cheng*. They mutually generate each other in an endless cycle, and a strategist has to make smart decisions based on the interaction of the opposite factors.

Principle 5: Deception, maneuvers and flexibility. According to Sun Tzu's advice, a general needs to conceal the army's capability, movement, location and morale status in order to confuse, fool and mislead the enemy. Maneuvering is an important approach to seek an opportunity by moving the army. A commander must be able to march by an indirect route and divert the enemy. Flexibility means that the commander adjusts his planning to changing circumstances. When the enemy concentrates, prepare against him; where he is strong, avoid him. Put the enemy under constant strain and stress to wear him down.

These principles represent the essence of Sun Tzu's strategic thinking. At a higher level, they constitute an essential part of the ancient Chinese strategic culture. As a matter of fact, these military terms and principles are the third element that helps to build up the canonical status of *AOW* in Chinese strategic culture.

4.1.3 Multiple paratexts

As mentioned in Section 1.1, *AOW* has been widely circulated in various editions and well received for generations since it was composed. A large school of Chinese generals, scholars, men of letters and politicians have deliberated on and applied Sun Tzu's strategic ideas. Some wrote down their comments

Table 4.2 Paratexts of ST

No.	Title in Chinese	Title in English	Time of composition	Chinese characters
1	十家注	Commentary by Chinese scholars	155–1279	109,388
2	孙子兵法序（孙星衍）	Preface by Sun Hsing-yen	1877	996
3	孙子序（曹操）	Preface by Ts'ao Ts'ao	155–220	202
4	遗说 (郑友贤)	I Shuo by Cheng You-Hsien	1127–1279	6,766
5	孙子本传 (司马迁)	Biography of Sun Tzu from *Shi Chi* by Ssu-ma Ch'ien	145 BC–90 BC	1,129
6	孙子叙录(毕以珣)	Sun Tzu Hsu Lu by Pi I-Hsün	1757–1836	8,565

and others authored prefaces. Later, these comments, prefaces and other relevant paratexts were collected by scholars into different editions of *AOW*.

The ST, *The Art of War Commented by Ten Scholars* edited by Sun Hsing-yen, who claimed to be a descendent of Sun Tzu, is an edition with multiple paratexts. It collects 6 paratexts, with more than 127,000 Chinese characters, as shown in Table 4.2. These paratexts, composed by different editors and commentators across from 145 BC to 1877 AD, have become a vital addition to Sun Tzu's military theory and an integral part of the ancient Chinese military culture.

1   Commentary by Chinese scholars. Among all paratexts, the commentary by Chinese scholars and generals is the largest in number of words, adding up to almost 20 times the size of the core text. Comments are put right after each sentence. Although the book is entitled *The Art of War Commented by Ten Scholars,* the comments are actually added by 11 generals and scholars, from the second to the nineteenth century. They are Ts'ao Ts'ao (曹操), Meng Shih (孟氏), Li Ch'üan (李筌), Chia Lin (贾林), Tu You (杜佑), Tu Mu (杜牧), Chen Hao (陈皞), Mei Yaoch'en (梅尧臣), Wang Hsi (王晳), Ho Shih (何氏) and Chang Yü (张预). Some of them are famous generals, such as Ts'ao Ts'ao (155–220), one of the greatest military geniuses, famed for the marvelous rapidity of his marches; while others are men of letters, such as Li Ch'üan, a well-known writer on military tactics of the eighth century, and Tu Mu (803–852), a great poet in the Tang Dynasty who was erudite in military history.

2   Preface by Sun Hsing-yen. Quoting extensively from Chinese historical records, this preface offers a traditional view of Sun Tzu's life and feats, and a brief introduction to different versions of *AOW*.

3   Preface by Ts'ao Ts'ao. Ts'ao Ts'ao is believed to be the first commentator on Sun Tzu. His preface gives a sketch of Sun Tzu's life and a commendatory comment on *AOW*.

4   I Shuo (遗说). This is a miscellany of Sun Tzu information by a scholar in the Sung Dynasty (960–1279). It includes an interpretation of Sun Tzu's military theory in aspects that have not been touched by former commentators.

5   Biography of Sun Tzu. Excerpted from the *Shih Chi* by Ssu-ma Ch'ien (145–90 BC), this biography features a detailed and vivid narration of Sun Tzu's legendary story of training concubines to be soldiers and winning the trust of King Ho Lü.

6   Sun Tzu Hsu Lu by Pi I-Hsün. Pi I-Hsün (1757–1836) was a reputed scholar in the Qing Dynasty. His essay consists of five parts. He firstly quoted from seven historical books to report the life story of Sun Tzu. Then he provided the interview between King Ho Lü and Sun Tzu. Thirdly, he pointed out the omissions and corruptions in 13 chapters. Fourthly, he made a list of quotes from Sun Tzu in other books, to illustrate the popularity of Sun Tzu in China. Finally, there he provided a briefing about different versions of *AOW.*

These paratexts include a variety of sources of information about the core text. They explain the meanings of words, phrases and sentences, substantiate Sun Tzu's military principles with famous battles in Chinese history and interpret significant cultural elements from ancient Chinese geography, language, social life, political systems, economics and so on. If we focus on the contribution of the paratexts to strategic culture, we can discern two approaches to Sun Tzu's text: supplementation of cultural information and evaluation.

Supplementation

The SL paratexts supply information about strategic culture in two aspects: firstly, the biography of Sun Tzu, and secondly the interpretation as well as substantiation of Sun Tzu's strategic thinking.

Authorship is an extremely important issue for such a military classic as *AOW.* If the text was written by an unknown layman, there will be doubts among readers about how much strategic truth is contained in the text. If the text was composed by a prestigious general, adroit on the battlefield, his work will be highly esteemed and thus widely circulated. For such a reason, four paratexts provide biographic information about Sun Tzu. The biographic messages in "Preface by Ts'ao Ts'ao" and "Preface by Sun Hsing-yen" are rather short, composed of only several sentences. But the "Biography of Sun Tzu" from the *Shih Chi* by Ssu-ma Ch'ien, is a very detailed and vivid description of the life story of Sun Tzu in more than 1,300 words. The paratext "Sun Tzu Hsu Lu" by Pi I-Hsün not only copies down the biography from the *Shih Chi*, but also adds biographical notes about Sun Tzu from other historical records, such as *A history of Wu and Yue Kingdom* (吴越春秋), *An Anthroponymy Book in China* (姓氏辨证书) and *A Chorography of Yue Kingdom* (越绝书)

for cross-reference. All the biographic information contained in these four paratexts work together to tell readers the life story of Sun Tzu.

According to "The Biography of Sun Tzu" and "The Biography of Wu Tsihsu" in *Shi Chi* by the renowned historian Ssu-ma Ch'ien, Sun Tzu was born around 500 BC in the late Spring and Autumn period as a contemporary of Confucius in the State of Chi, with the full name 孙武 (Sun Wu) and courtesy name 长卿 (Changqing). He is commonly known as 孙子 (Sun Tzu) because 子 (Tzu) in ancient Chinese is the honorific title for a learned or virtuous man. Sun Wu authored his book of 13 chapters on war, and was recommended by General Wu Tsuhsu to King Ho Lü of the State of Wu. Having read the book, King Ho Lü was impressed by the strategist's acute intelligence in military affairs. The king then tested Sun Tzu's skills and theories by asking him to train a group of 180 concubines into soldiers. Sun Tzu took the challenge. He divided these ladies into two companies and appointed the king's two favorite concubines as the heads of the two companies. When the concubines were ordered to practice some simple drills such as "face right" and "left turn", they only giggled. Sun Tzu said that if the orders were not clearly sent out, the general should be responsible. He then reiterated the command, and the concubines burst into laughter again. Sun Tzu said if the orders were clear and the commands were not followed, it was the soldiers to blame. He firmly ordered the execution of the companies' two heads, against the king's protests. Sun Tzu insisted that once a general was appointed, it was his duty to carry out the mission. After the two leaders of the company were executed, another two officers were chosen. Soon, the concubines were strictly drilled and disciplined, ready to fight under any circumstances. Seeing the great skills that Sun Tzu had in leading an army, King Ho Lü appointed him general. In the west of Wu, Sun Tzu defeated a superior power, the state of Chu, with an army of 30,000 against the enemy's 200,000. He also forced his way into the capital of state of Ying. To the north of his country, Sun Tzu intimidated states of Chi and Chin. His fame spread far and wide among other feudal lords. A hundred years after Sun Tzu's death, one of his descendants, Sun Pin, also became a military genius, and composed another book on war.

This legendary story of Sun Tzu training concubines into soldiers is extremely important. On one hand, it narrates the authorship of the text. On the other hand, it portrays some important strategic truths about how to train and handle an army with orders and awe. Such a deed exemplifies some conditions under which victory can be achieved, as pointed by Sun Tzu in chapter 1 of *AOW*: the general is capable; the discipline is rigorously enforced, the army is strong, the officers and men are highly trained and both reward and punishment are constantly practiced in the army. Sun Tzu's life story also coincides with what he says in chapter 10: "if you are indulgent and incapable, your soldiers must be likened to spoilt children and useless for any practical purpose". Therefore, the life story of Sun Tzu adds some persuasive power to the strategic thinking contained in the core text.

Another important role of the ST paratexts is that they provide explanation as well as substantiation of Sun Tzu's strategic thinking. According to our survey, the core text of 6,100 Chinese characters in 594 sentences is interpreted and substantiated by the Chinese comments of about 109,300 characters in 2,766 notes. This means that, on average, each Chinese character in the core text is interpreted by 18 characters and each Chinese sentence is commented on by about 4.6 notes.

The explanation of the core text is necessary for contemporary and future readers because *AOW* is versed in pithy and profound ancient Chinese language. Substantiation is also essential because Sun Tzu's strategic thinking, expressed in short, powerful and figurative statements, lacks concrete evidence from the battles and military leaders in history. What the comments do exactly is to make up for the above two deficiencies of the core text and generate easier comprehensibility for future readers. Cheng You-Hsien remarked in his preface to I Shuo that:

> without the comments, students of *AOW* can find no access into the science of strategy, nor trace its origin and development, nor gain the mastery of military science. These comments are undoubtedly a huge aid to *The Art of War*.
>
> (而学兵之徒, 非十家之说, 亦不能窥武之藩篱; 寻流而之源, 由径而入户, 于武之法, 不可谓无功矣)

For a better understanding of the paratexts' role, we resort to chapter 1 as an example. The core text of chapter 1 contains only 42 sentences in 427 Chinese characters. However, there are 221 comments in 13,315 characters. Among them, 131 comments in 8,130 characters expound on the meaning of the core text, while 90 comments in 4,885 characters provide concrete examples from battles and generals in ancient China.

If we take into consideration the intensive use of military terms in chapter 1, as shown in Table 4.3, we can see that the commentators provided information on battles and generals to help substantiate Sun Tzu's military thoughts. In the core text of chapter 1, although 12 military terms in strategy and tactics are included, there is no discussion of battles and officers. However, in the comments to chapter 1, battles and generals are mentioned 231 times,

Table 4.3 Military terms in chapter 1 of ST

Category	Terms in Core Text		Terms in Comments	
	Type	*Frequency*	*Type*	*Frequency*
Strategy and tactics	12	35	15	377
Battles and officers	0	0	146	231

together with 146 terms, which adds greatly to the persuasive power of Sun Tzu's pithy strategic statements. Furthermore, these comments constitute an encyclopedic body of military ideas, events and people, which are particularly important in the representation of ancient Chinese strategic culture and in establishing the canonical status of *AOW*.

To further understand how the comments work in partnership with the core text, an example from chapter 1 is quoted below. For the sake of convenience, the comments are numbered respectively, and a gloss is provided by the present author.

Example 1
Core Text: 兵者，诡道也。(Warfare is based on deception.)

Comments: (1) 曹公曰：兵无常形，以诡诈为道。(2) 杜牧曰：兵无常形，以诡诈为道。若息侯诱蔡，楚子谋宋也。(3)李筌曰：军不厌诈。(4) 梅尧臣曰：非谲不可以行权，非权不可以制敌。(5) 王晳曰：诡者，所以求胜敌；御众必以信也。(6)张预曰：用兵虽本于仁义，然其取胜必在诡诈。故曳柴扬尘，栾枝之谲也；万弩齐发，孙膑之奇也；千牛俱奔，田单之权也；囊沙壅水，淮阴之诈也。此皆用诡道而制胜也。

Gloss: (1) Ts'ao Kung: In war, there is no constant condition. Therefore, all warfare is based on deception. (2) Tu Mu: In war, there is no constant shape. War is fought on the basis of deception. This is the strategy used when the **Marquess of Xi** was luring the **Marquess of Cai Ai** into trap, and when **Prince of Chu** was attacking the s**tate of Song**. (3) Li Ch'üan: There can never be too much deception in war. (4) Mei Yaochen: Nothing but deception can lead to a stratagem. Nothing but a stratagem can lead to victory. (5) Wang Hsi: Deception is the means to gain victory against an enemy; however, to lead an army one must depend on trust. (6) Chang Yu: War is fought for the purpose of benevolence and justice; however, it is won by means of deception. Thus, **Luan Tsi,** ordered his soldiers to attach tree branches to chariots to raise dust, feinted a rout and won the war against the enemy. **Sun Pin** deceived Pang Juan into believing that the Qi army was in a state of cowardice and finally defeated Pang Juan. **Tian Tan** pretended that his besieged city was ready to surrender and eventually drove away the enemy with the help of a thousand oxen. **Han Hsin** and his army dammed a river with sandbags, drawing his enemy downstream, and crushed them with the released currents. All these generals employed deception and trickery for the sake of triumph. [The gloss was translated by the present author. The numbers were added to mark the comments by different Chinese scholars.]

In Example 1, the core text is a concise statement of only five Chinese characters, announcing one of the key military principles proposed by Sun Tzu: deception. Its profundity in truth has attracted many scholars and generals to interpret and exemplify it with cases in point. There are six comments by six scholars using 168 characters. Firstly, these comments work together to explain Sun Tzu's thoughts on deception from different perspectives, although some are repetitive. The first comment by Ts'ao Kung (an honorable title for Ts'ao Ts'ao) and the second by Tu Mu explain why deception is needed. The third comment is a reiteration of Sun Tzu's statement in the core text. The fourth to sixth comments explain the purpose of deception.

Secondly, these comments substantiate Sun Tzu's strategic thinking with a list of six famous generals and six battles led by them in Chinese history: Marquises of Xi, Prince of Chu, Luan Tsi, Sun Pin, Tian Tan and Han Hsin. These military anecdotes offer concrete evidence for the principle of deception and add to the persuasive power of Sun Tzu's strategic statement. Meanwhile, they constitute a comprehensive representation of the principle of deception, including its concept, the battles applying this principle and the people who were skilled in deception.

Evaluation

In addition to supplementation, evaluation is another approach used in the paratexts' representation of Sun Tzu's strategic culture. In the SL paratexts, three authors from different dynasties offered their evaluations of Sun Tzu and *AOW*: Ts'ao, Cheng You-Hsien and Sun Hsing-yen.

In a paratext entitled "Preface", Ts'ao Ts'ao sang high praise of *AOW*: "Of all the books I have read on war and fighting, Sun Wu's is the profoundest" (吾观兵书、战策多矣, 孙武所著深矣). Ts'ao Ts'ao was one of the greatest military geniuses in China, famed for his comprehensive calculations before battles. He was also the king of the Wei Kingdom and was addressed honorifically as Wei Wudi (魏武帝), which means the King of Military Prowess. A commendatory comment by such a great general as Ts'ao Ts'ao certainly added credit to the prestigious status of *AOW*. Ts'ao Ts'ao also remarked that "in military terms, the Sage's rule is normally to keep the peace and to move his forces only when occasion requires. He will not use armed force unless driven to it by necessity" (圣人之用兵, 戢而时动, 不得已而用之). This evaluation highlighted the peace orientation of *AOW*.

Cheng You-Hsien, in his "Preface to I Shuo", acclaimed Sun Tzu's military thoughts: "In the eyes of the strategist, it is a book of strategy. For those who are smart, it is a book of wisdom. Soldiers in the army are told to follow the book but they don't understand why" (是以谋者见之谓之谋, 巧者见之谓之巧, 三军由之而莫能知之). Chen's evaluative remarks point out that Sun Tzu's work can be read as a book of strategy and also as a book of wisdom in a general sense.

Sun Hsing-yen, the editor of the ST, highly acclaimed Sun Tzu in his own preface:

> His book is deep in Three Geniuses of Heaven, Earth and Men, as well as The Five elements of metal, wood, water, fire and earth. Based on benevolence

and righteousness, aided by strategy and tactics, *AOW* is just and true indeed. Ancient generals who counseled and acted upon it conquered, those who broke its rule and acted against it were defeated. Thus, *AOW* is reputed as a military canon. When compared with Six Classical Arts: rites, music, archery, riding, writing, arithmetic, *AOW* is definitely worthy of its fame.
（其书通三才、五行，本之仁义，佐以权谋，其说甚正。古之名将用之
则胜，违之则败，称为"兵经"。比于六艺，良不愧也）

Sun Hsing-yen used words and phrases like "deep", "benevolence and right-eousness", "just and true", "a military canon" and "worthy of its fame" to establish the respectable status of *AOW*. In conclusion, Sun Hsing-yen remarked that: "Sun Tzu's 13 chapters is a must read for people" （则十三篇
之不可不观也）.

The evaluative remarks by three authors in the SL paratexts are helpful in establishing a positive image of Sun Tzu and his book, and consequently encourage readers to appreciate the military thinking.

To summarize, as we have discovered in the above section, the SL core text features profound, unique military concepts and principles with a succinct and vivid style of representation. However, such profundity and succinctness may turn out to be a challenge for modern readers and translators.

Fortunately, a huge body of paratexts compensates for the core text by supplementing additional cultural information, such as Sun Tzu's life story and historical battles, to interpret and substantiate Sun Tzu's strategic ideas. The paratexts also provide evaluative remarks, enthusiastically praising Sun Tzu's military wisdom.

4.2 Strategic culture reconstruction in Giles' translation

In the following section, we will focus on TT1 to find out how Sun Tzu's strategic concepts and principles have been translated in the core text and how the paratexts contribute to the reconstruction of Chinese strategic culture.

4.2.1 *Translation of the core text*

Three questions will be answered when examining how Sun Tzu's strategic thinking is translated in the core text: (1) What methods and approaches are employed in translating military terms? (2) How are the most important strategic principles translated? and (3) How is the language style handled? It is hoped that the answers to these questions will provide a clear picture of the translation of Sun Tzu's strategic thoughts.

To help address the above questions, a sentence-to-sentence survey was done to investigate whether the core text is fully translated based on the parallel bilingual corpus. Giles' division of ST sentences is slightly different from the division by Sun Xinyan, and consequently there are 385 sentences in Giles' version. It was found that almost all the sentences from ST were translated and that omission of sentences and their constituents was rare. As a result, it can be said that the core text of TT1 is a full translation of ST.

Translation of military terms

To secure an overview of how the ancient Chinese military terms were translated in the core text, another survey was conducted about the methods and approaches used in translating. The result of the survey is shown in Table 4.4. In TT1 about 83% of military terms were translated with the domesticating approach, 17% with the foreignizing approach. Since the use of domestication far outweighs that of foreignization, it is clear that Sun Tzu's strategic concepts have been translated with a focus on readability and accessibility. This domesticating approach reduces readers' difficulty in understanding. However, it has to be noted that, at the same time, such an approach smothers the exotic flavors and irons out the unique qualities of these terms.

Under the domestication approach, paraphrasing is the most frequently used method, with more than 70% of military terms translated into words and phrases that are culture-neutral. For instance, the term 夺气 is translated into "be robbed of its spirit", and 夺心 into "be robbed of his presence of mind". 气 is a polysemous and important word extensively used in Chinese Taoist philosophy, traditional medicine and many other fields. The original meaning of 心 is "heart", with extended meaning of "mind", "emotion" or "soul". Although the translations of 气 and 心 as "spirit" and "mind" are easy to understand, they fail to represent the complex set of subtle meanings conveyed by them. Another term, 轒辒, refers to military equipment used in ancient China to besiege a walled city. It is explained rather plainly in TT1 as "movable shelters". For ordinary readers, such a translation is accessible; however, it may not be sufficient for military experts who want to further understand the component and usage of such equipment.

Substitution, another domesticating method used to translate the Chinese military terms, increases their intelligibility and accessibility, but at the cost

Table 4.4 Translation approaches and methods for military terms in TT1

	Foreignization		Domestication		
	Calque	*Transliteration*	*Paraphrasing*	*Generalization*	*Substitution*
Strategy and tactics	13	0	106	0	3
Supply and armament	4	0	18	1	5
Terrain	4	1	21	0	0
Organization	2	0	10	0	5
Battles and officers	1	7	0	0	0
Total 204 (100%)	24 (12%)	10 (5%)	156 (76%)	1 (1%)	13 (6%)

of Chinese cultural essence. For instance, 钟, a Chinese measure of volume, is replaced by "cartload", a Western measurement. Target readers are deprived of the opportunity to acquaint themselves with this unique ancient Chinese measure. Another phrase, 廊庙, is rendered in TT1 as "council-chamber". 廊庙 refers to a great temple where the Chinese king discusses national affairs with his ministers, and it is quite different from a council in a Western political system. Therefore, the translation "council-chamber" needs further explanation; otherwise, the difference between the Chinese temple and the English council-chamber will be eradicated.

A relatively small number of military terms in TT1 are foreignized, aiming at retaining the exotic flavor of Sun Tzu's military concepts, and the brevity of style of the ST is maintained in the core text at the same time. Therefore, there comes a need to reduce the difficulty in understanding these ancient Chinese military terms for target readers who are usually familiar with Western military terms.

Calque, a foreignizing method, is used to translate some military terms in TT1 word by word. For example, 轻车, a kind of ancient Chinese military equipment, is calqued as "light chariots" in TT1. This rendition, however, fails to inform readers of the size and usage of the vehicles on a Chinese battlefield. 庙算 is an important planning activity which involved an ancient Chinese army's use of divination or numerology prior to fighting. It is rendered word-by-word as "calculations in his temple" in TT1, which means that military officers were planning in the temple before battle. Such a translation ignores the difference between "calculations" and "divination" or "numerology", and consequently does not fully convey the meaning of the term. Some Chinese abbreviations are also translated word by word. For example, 五行 is translated as "five elements", 黄帝 as "the Yellow Emperor", 四帝 as "four sovereigns". These abbreviations were ancient terms rich in Chinese cultural essence and their short calquing translations do not provide enough cultural information. Readers without profound knowledge of ancient Chinese history tend to be puzzled by these abbreviated phrases in TT1.

Other military terms are transliterated in TT1 so that the Chinese flavor is preserved. The names of ancient generals, strategists and kings are in most cases translated according to their pronunciation. For instance, 伊挚 is rendered as "I Chih"; 吕牙 as "Lü Ya"; 刿 as "Kuei"; 诸 as "Chu". However, these translations may turn out to be strange and obscure terms if no extra help is provided for ordinary Western readers who are not well-versed in Chinese history.

The above analysis of the translation of military terms reveals that neither foreignization nor domestication alone can fully convey the cultural connotations of military terms in the core text. On one hand, although most of the terms are domesticated so that readability and accessibility is accentuated, additional help is still needed to bring out the terms' exotic cultural essence to the fullest extent. On the other hand, when some terms are foreignized, extra information is still needed to ensure their intelligibility.

Translation of strategic principles

Now, we will look at how the key principles of Sun Tzu's strategy are translated in TT1. These principles are expressed in a bunch of sentences among the 13 chapters and some sentences are much more important than others. Sentences that best illustrate each of these five principles have been used for the example analysis.

Example 2 Victory without fighting (chapter 3)

ST: (1) 凡用兵之法，全国为上，破国次之；全军为上，破军次之；全旅为上，破旅次之；全卒为上，破卒次之；全伍为上，破伍次之。(2) 是故百战百胜，非善之善也：不战而屈人之兵，善之善者也。(3) 故上兵伐谋，其次伐交，其次伐兵，其下攻城。(4) 攻城之法为不得已。

TT1: (1) In the practical art of war, the best thing of all is to take the enemy's country whole and intact; to shatter and destroy it is not so good. So, too, it is better to capture an army entire than to destroy it, and to capture a regiment, a detachment or a company entire rather than to destroy them. (2) Hence to fight and conquer in all your battles is not supreme excellence; supreme excellence consists in breaking the enemy's resistance without fighting. (3) Thus the highest form of generalship is to baulk the enemy's plans; the next best is to prevent the junction of the enemy's forces; the next in order is to attack the enemy's army in the field; and the worst policy of all is to besiege walled cities. (4) The rule is not to besiege walled cities if it can possibly be avoided. [The numbers were added by the present author.]

Example 2 illustrates the translation of Sun Tzu's principle of victory without fighting and his dislike of militarism. In the first sentence, there are four terms for military organization in ancient China, 军, 旅, 卒 and 伍. Each of these four terms refers to a military unit, for about 10,000 soldiers, 500, 100 and 5 respectively, according to Chinese military theorists and Sun Tzu experts such as Guo Huaruo (1984, 99) and Li Ling (1991, 14). There are two options for translators: to foreignize or domesticate them. In TT1, these terms are translated as "an army", "a regiment", "a detachment" and "a company" respectively. In the British system of military units, an army consists of 100,000 soldiers; a regiment, 1000 to 2000; a company, 100 to 250. TT1 employs the approach of domestication, to an effect that text readability is prioritized. However, the numerical differences between the English terms and Chinese ones are ignored, resulting in a loss of the essence of ancient Chinese military culture.

The phrase 百战百胜 in the second sentence is paraphrased as "to fight and conquer in all your battles". 百 in Chinese can either refer to the precise number, "one hundred", or simply suggest "a large number", and therefore, the translation has failed to convey the first meaning.

The third sentence is of paramount importance in Sun Tzu's thesis, according to Li (1991, 15) and Guo (1984, 101). This sentence lists the four types of warfare in a hierarchic order from the best to the worst: 伐谋, which means the game of strategy before the war breaks out; 伐交, which means the wrestling of diplomatic efforts or the rupture of a coalition; 伐兵, which refers to the battle in field; and 攻城, which suggests the attack against a walled city. There is also a scale of increasing intensity among the four, as to win by strategy is the least costly and bloody, and to besiege cities is the most costly and bloody. However, in TT1, 谋 is translated as an ordinary word "plan" and the translation fails to indicate that it is the highest order of strategy among the four forms of warfare. "伐交" is improperly translated as "to prevent the junction of the enemy's forces", which breaks down the stratified order and fails to highlight the importance of diplomatic efforts. The translations of these two terms fail to highlight the non-military means before the war starts, which Sun Tzu usually thinks highly of in his book.

According to the above analysis, the translation in this example manages to interpret Sun Tzu's strategic ideas but not without mistakes. It renders the general meaning of his principle but fails to fully bring subtle linguistic and cultural nuances into English.

Example 3 Holistic approach (chapter 2)

ST: (1) 其用战也贵胜，久则钝兵挫锐，攻城则力屈。(2) 久暴师则国用不足... (3) 故兵闻拙速，未睹巧之久也。(4) 夫兵久而国利者，未之有也。

TT1: (1) When you engage in actual fighting, if victory is long in coming, then men's weapons will grow dull and their ardour will be damped. If you lay siege to a town, you will exhaust your strength. (2) Again, if the campaign is protracted, the resources of the State will not be equal to the strain ... (3) Thus, though we have heard of stupid haste in war, cleverness has never been seen associated with long delays. (4) There is no instance of a country having benefited from prolonged warfare.

Example 3 illustrates Sun Tzu's holistic approach to war. He not only outlines the tactics on battlefields, but also perceives other factors that may be at play in strategy, such as 兵 and 力, the physical power of an army, 锐, psychological factors and 国用, economic factors. Generally speaking, TT1 has rather faithfully and gracefully translated sentences 2 and 4. For instance, 钝兵, is properly translated into "weapons become dull", 挫锐 into "ardour will be damped", 不足 into "not be equal to the strain". These English words are well-chosen and fit in the context to illustrate the physical, psychological and economic elements in Sun Tzu's holistic approach to strategy.

However, the translation still has its shortcomings. When readers come across "stupid haste", the word-by-word translation for 拙速, in the third sentence of the ST, obscurity may cloud their mind. The translation suggests that haste is stupid, or swiftness is blundering, which is contradictory to what Sun

Tzu preaches for the quick ending of war. It can easily lead to readers' confusion if no further explanation is furnished. Actually, 拙 means "plain" or "simple" and 拙速 means "plain swiftness".

Example 4 Foreknowledge and information (chapter 3)

ST: 故曰：知己知彼，百战不殆；不知彼而知己，一胜一负；不知彼不知己，每战必殆。

TT1: Hence the saying: If you know the enemy and know yourself, you need not fear the result of a hundred battles. If you know yourself but not the enemy, for every victory gained you will also suffer a defeat. If you know neither the enemy nor yourself, you will succumb in every battle.

In this example, Sun Tzu puts forward a universal law of war: adequate information of the enemy and oneself will ensure victory. To add to the persuasive power, three situations of different degrees of information are listed in parallel structure. Therefore, the parallel structure is essential in representing Sun Tzu's military ideas and it is faithfully maintained in TT1. The word 殆, which means "in danger", is not clearly stated in the first sentence, but smartly rendered into "succumb" in the third sentence.

Example 5 Dialectical view (chapter 5)

ST: 三军之众，可使必受敌而无败，奇正是也。兵之所加，如以破投卵者，虚实是也。凡战者，以正合，以奇胜。故善出奇者，无穷如天地，不竭如江河。

TT1: To ensure that your whole host may withstand the brunt of the enemy's attack and remain unshaken—this is effected by **manoeuvres direct and indirect**. That the impact of your army may be like a grindstone dashed against an egg—this is effected by the science of **weak points and strong**. In all fighting, the **direct method** may be used for joining battle, but **indirect method**s will be needed in order to secure victory. **Indirect tactics**, efficiently applied, are inexhaustible as Heaven and Earth, unending as the flow of rivers and streams. [The emphasis of words has been added by the author of this book.]

Example 5 expounds on two pairs of concepts, 奇 and 正, 虚 and 实，which reflect the dialectical view taken by Sun Tzu. 奇 (*ch'i*) and 正 (*chéng*) is a pair of Sun Tzu's fundamental strategic concepts (O'Dowd and Waldron 1991, 30). The *chéng* tactics include employing troops in direct, conventional or expected ways, such as massive frontal assaults; while the *ch'i* tactics are realized through employing flexible forces in indirect, unconventional and unexpected ways, such as flanking and rear attacks. There is also an interplay and conversion between *ch'i* and *chéng*. In some cases, *chéng* can turn into *ch'i* and vice versa, and they mutually produce each other in an endless cycle. These primarily dialectical concepts originate from the Taoist philosophical view, and are well-known by the Chinese audience. However, due to the lack of definitions or other explanations in the core text, these profound and polysemous concepts may seem alien and difficult for English readers.

TT1 opts for domesticating translations for certain key concepts. For instance, 正 is interpreted as "direct" in the phrases "manoeuvres direct" and "direct method"; 奇 is translated as "indirect" in "manoeuvres indirect", "indirect methods" and "indirect tactics". Multiple phrases are better than a single expression in that they help to convey the profundity of this strategic principle. However, TT1 provides no further illustrations in the core text. Most readers may still be confused about how 奇 and 正 can be used interchangeably.

Another pair of concepts, 虚 and 实, also play an important part in Sun Tzu's dialectical view of warfare, and 虚实 is used as the title of chapter 6. In Chinese, 虚 may mean emptiness, weakness, void, inanity, incapability or falsity; while 实 may mean fullness, strength, solid, capability or truth. In TT1, 虚 and 实 are interpreted as "weak points and strong". For these important terms, neither definitions nor concrete examples are provided in the core text to illustrate the multiple and profound meanings.

Similes are used when Sun Tzu is discussing these two pairs of concepts. In sentence 2, 以实击虚 (to attack void with solid) is compared to a grindstone striking an egg. The multiple application of 奇 and 正 is viewed as inexhaustible and as unending as the flow of rivers and Heaven and Earth. TT1 faithfully translates these similes, which is helpful in transmitting Sun Tzu's mode of strategic thinking and retaining the language style of ST.

Generally, the translation of strategic principles in Example 4 shows that the interpretation method used in the core text, despite its advantages in terms of intelligibility and conciseness, has limitations in terms of conveying the multiple and profound meanings of Sun Tzu's military principles.

Example 6 Deception, maneuverability and flexibility (chapter 1)

ST: (1) 兵者，诡道也。 (2) 故能而示之不能，用而示之不用。 (3) 近而示之远，远而示之近。 (4) 利而诱之，乱而取之。 (5) 实而备之，强而避之。 (6) 怒而挠之，卑而骄之。 (7) 佚而劳之，亲而离之。 (8) 攻其无备，出其不意。 (9) 此兵家之胜，不可先传也。

TT1: (1) All warfare is based on deception. (2) Hence, when able to attack, we must seem unable; when using our forces, we must seem inactive; (3) when we are near, we must make the enemy believe we are far away; when far away, we must make him believe we are near. (4) Hold out baits to entice the enemy. Feign disorder, and crush him. (5) If he is secure at all points, be prepared for him. If he is in superior strength, evade him. (6) If your opponent is of choleric temper, seek to irritate him. Pretend to be weak, that he may grow arrogant. (7) If he is taking his ease, give him no rest. If his forces are united, separate them. (8) Attack him where he is unprepared, appear where you are not expected. (9) These military devices, leading to victory, must not be divulged beforehand.

In Example 6, Sun Tzu declares another key principle in strategy: deception, maneuver and flexibility. With 7 sentences (from second to eighth) in parallel structures, Sun Tzu specifies a dozen ways in which this principle can be applied, including feigning inability and inactivity, covering up location, luring the enemy with baits, irritating and harassing the enemy, seeding arrogance or disorder among the opponent and launching a surprise attack. These methods are the components of the rule of disguise, flexibility and surprise attack. 诡道, in Sun Tzu's mind, is a neutral and comprehensive concept, of which deception is just an ingredient (Guo 1984, 88). However, TT1 translates 诡道 as a derogatory term, "deception", which rules out other components of the concept.

These seven sentences also point to a feature of ancient Chinese language: omission of subjects for the sake of brevity. Such terseness can lead to multiple interpretations since different potential subjects can be added. For instance, the fourth sentence can take different subjects and yield multiple acceptable explanations (Guo 1984, 84–85). With 敌, which means "enemy", added as the subject, the sentence 利而诱之，乱而取之 can turn into （敌）利而（我）诱之，（敌）乱而（我）取之. This means that if the enemy lusts for bait, we may lure him; if the enemy is in disorder, we can strike him. Another solution is that 我 is added as the subject: （我）利（敌）而诱之，（我）乱（敌）而取之. In this case, the sentence means that we can lure the enemy with bait, sow disorder among the enemy and strike him. Similarly, the sentence 怒而挠之，卑而骄之 may convey two distinct meanings if different subjects are added. It may become （敌）怒而（我）挠之，（敌）卑而（我）骄之, meaning that if the enemy is angry, we can harass him; if he is humble, we can make him arrogant. It can also be turned into （我）怒（敌）而挠之，（我）卑而（敌）骄之, which means that we can make the enemy angry and irritate him; we can pretend to be inferior and encourage the enemy's arrogance. However, in TT1, the meanings of the sentences are nailed down, ruling out other potential interpretations.

Another issue to be considered in Example 6 is that Sun Tzu, as he often does elsewhere, simply declares his principle and notes down his guidelines for its application. Then he stops there and doesn't supply concrete examples of battles to illustrate, testify to or substantiate the principle.

The above analysis of the translations of five principles shows that Sun Tzu's strategic ideas are on the whole gracefully conveyed in TT1. Firstly, with most of the military terms paraphrased so as to be easily intelligible and highly readable, the translations are crystal clear and accessible. However, it has to be noted that Sun Tzu's exotic flavor is not fully retained but is lost to some extent since the method of paraphrasing in English terms is most frequently used. Secondly, as the succinct style is maintained in TT1, the profound and multiple meanings of military terms and principles are not carried over to the fullest extent. Thirdly, the lack of definition and substantiation of the abstract principles is also closely inherited in TT1. Such sacrifice of foreign flavor, profundity and multiple explanations, together with the lack of substantiation, is believed to be the translators' lesser-of-two-evils choice. Our major concern here is whether these deficiencies can be compensated in the paratexts, which will be discussed in Section 4.2.

Translation of metaphoric expressions

As mentioned in Section 4.1, metaphoric expressions are essential in the representation of Sun Tzu's military theory. The translation of metaphors and similes is therefore significant in the reconstruction of the ancient Chinese strategic culture.

In the ST, similes and metaphors from Chinese culture reflect the distinct Chinese way of thinking, especially the Chinese mode of argumentation. By way of analogy, they establish links between the tenors and vehicles, specifically, between military concepts and other entities. If the analogical link is preserved and the vehicles retained in translation, the foreignization approach has been used. When the metaphoric expressions are replaced by target cultural counterparts or the meaning is paraphrased in plain words, such a translation can be classified as domestication.

According to our survey, 89% of the Chinese metaphors and similes are preserved in TT1, which means that foreignization is the most frequently used approach (see Table 4.5).

Example 7 Simile

ST: 夫兵形象水，水之形避高而趋下；兵之形，避实而击虚。水因地而制流，兵应敌而制胜。故兵无常势，水无常形。

TT1: Military tactics are like unto water; for water in its natural course runs away from high places and hastens downwards. So in war, the way is to avoid what is strong and to strike at what is weak. Water shapes its course according to the nature of the ground over which it flows; the soldier works out his victory in relation to the foe whom he is facing. Therefore, just as water retains no constant shape, so in warfare there are no constant conditions.

水(Water) is a word which frequently occurs (17 times) in Sun Tzu's text, with two usages: one referring to the natural object, "river", the other in similes. In Example 7, the maneuver of an army is compared to the flowing movement of

Table 4.5 Translation of metaphors and similes in TT1

	Total	Foreignization	Domestication
Simile	21	19 (90%)	2 (10%)
Metaphor	6	4 (67%)	2 (33%)
Overall	27	23 (89%)	4 (11%)

water, and Sun Tzu vividly explains the abstract notion of avoiding strength and attacking weakness. This analogical mode of thinking is rather efficient, highly concise and memorable though not necessarily strictly logical. In translation, the water simile is foreignized, preserving the figurative expression and informing the readers of the Chinese way of military thinking. As water flowing is universally seen, readers have easy access to the gist of the simile. Thus, the foreignizing translation is sufficient to convey Sun Tzu's mode of strategic thinking and the style of text.

Example 8 Metaphor

ST: 践墨随敌，以决战事。

TT1: Walk in the path defined by the rule and accommodate yourself to the enemy until you can fight a decisive battle.

In Example 8, 墨 means "ink", and 践墨 suggests that a carpenter in ancient China marked where he would cut through the wood with a thread saturated with ink. In the ST, 践墨 is a metaphorical expression for "acting according to rules" of military science. In TT1, the metaphor is paraphrased as "walk in the path defined by rule", which sacrifices the exotic flavor and vividness—most likely for the sake of economy in the core text.

The above analysis answers the three questions raised at the beginning of this section. (1) Generally speaking, for the translation of military terms, domestication is the most frequently used approach and paraphrasing is the most often applied method, which means that TT1 chooses to stress the accessibility of the core text. (2) The five strategic principles are translated in an accessible and graceful fashion, but not without shortcomings. (3) TT1 chooses to retain the concise and vivid style of the ST by approach of foreignization. However, as our analysis has shown, the readability of the text and the accessibility of Chinese strategic culture is somewhat compromised for the sake of target readers.

4.2.2 Role of paratexts

As the analysis in Section 4.1.2 has shown, paratexts are an essential part of ST. In order to get a whole picture of how the paratexts are dealt with in TT1, a list of the target paratexts has been made, as shown in Table 4.6.

The first part of the paratexts is Translator's Notes, with 586 items in over 51,300 English words. These notes fall into several categories: (1) selected translation or summarization of the ST comments in about 400 items and 46,200 words; (2) Giles' newly added references to famous Western generals, military events and books in 53 items and 2,600 words; (3) Giles' evaluation of Sun Tzu's military ideas in about 24 notes; (4) Giles' criticism of Calthrop's translation in 109 notes; and (5) other issues, such as the spelling differences of some Chinese words. The notes in the first three categories will be discussed in our study as they are more related to the reconstruction of strategic culture.

Table 4.6 Paratexts of TT1

Title	Sub-title	Relations with ST	Words
Translator's notes		Chinese comments translated or rewritten from the ST (46,233 words); Giles' newly composed notes on Western military culture (2,600 words), of evaluation on Sun Tzu, and criticism against Calthrop's translation (2,556 words)	51,389
Preface		Authored by Giles	1,009
Introduction	(1) Sun Wu and His Book	Partially written by Giles and partially translated from the ST paratexts. A biography of Sun Tzu by *Shi Chi* and Preface by Ts'ao Ts'ao translated. Preface by Sun Hsing-yen and Sun Tzu Hsu Lu by Pi I-Hsün partially translated	5,821
	(2) The Text of Sun Tzu	Partially written by Giles, with Preface by Sun Hsing-yen partially translated	1,087
	(3) The Commentators	Added	2,300
	(4) Appreciations of Sun Tzu	Added	379
	(5) Apologies for War	Partially written by Giles, with Preface by Sun Hsing-yen partially translated	1,791
	(6) Bibliography	Added	981
Total			64, 757

The Preface, a new piece added by Giles, explains why he decided to retranslate *AOW* and points out the features of his own new translation. In Giles' (1910, vii) view, the French translation by Père Amoit in 1782 is "little better than an imposture", containing "a great deal that Sun Tzu did not write, and very little indeed of what he did". The first English translations by Captain Calthrop in 1905, based on the Japanese text, is plagued by frequent omissions, willful distortion, and a lack of Chinese commentary. Giles believed that Sun Tzu deserved a better fate. He also lists a few special features in the translation: the division of the text into numbered paragraphs; a complete concordance of Chinese characters; the presence of text, translation and notes on the same page, following the Chinese ST layout; and a portion of the massive body of Chinese comments.

The Introduction, newly added by Giles, is a comprehensive piece of literature composed of six pieces: "Sun Wu and His Book", "The Text of Sun Tzu", "The Commentators", "Appreciations of Sun Tzu", "Apologies for War" and

"Bibliography". (1) "Sun Tzu and His Book" traces the legendary life of Sun Tzu, and the date of composition of *AOW*. (2) "The Text of Sun Tzu" gives a summary of different versions of *AOW*, from its appearance till the Qing Dynasty. (3) "The Commentators" is added by Giles to introduce the eleven Chinese commentators, their life stories, their military feats and the features or quality of their comments. (4) "Appreciations of Sun Tzu" illustrates the potent fascination with Sun Tzu by the greatest men of China, and lists the famous generals known to have studied Sun Tzu. (5) "Apologies for War" states the views on warfare in ancient China. It begins with the statement that China is a country that has experienced wars of all phases and ceaseless clashes of arms, and thus boasts a succession of illustrious generals, such as Po Ch'i, Ts'ao Ts'ao, Li Shih-min and Li Ching. However, Giles stressed, the major Chinese sentiment has been "consistently pacific and intensely op-posed to militarism in any form" (Giles 1910, xliv). In order to reflect this unorthodox view on war, Giles also collects and translates a few passages from Ssu-ma Ch'ien, Tu Mu and Chu His (朱熹) to highlight pacifism. Some parts of Sun Hsing-yen's Preface from the ST are translated. (6) The "Bibliography" introduces some ancient Chinese treaties on war after Sun Tzu and lists some Chinese encyclopedias as well as some historical works containing sections devoted to war.

The above brief introduction of the paratexts shows that Giles is fully aware of the significance of the paratexts in transmitting Chinese culture to his readers. He criticizes Calthrop for the neglect of "a much more numerous and infinitely more important 'army' of Chinese commentators" (Giles 1910, ix), and he translates some paratexts into TT1. He also tries to add the paratexts he considers necessary, such as the Introduction. Since there is such a huge mass of comments and lots of them are repetitive, Giles selectively translates and sometimes rewrites them, aiming to "extract the cream only" (Giles 1910, ix).

In the following part of this section, we will investigate how TT1 paratexts contribute to the reconstruction of ancient Chinese strategic cul-ture by means of supplementation, comparison and evaluation, taking into consideration both the original paratexts from ST and the new ones added by Giles.

Supplementation

As mentioned above, supplementation to the SL core text is helpful in building up the status of *AOW* as a military classic. Native readers of *AOW* may find the core text less difficult as they know who the author is, when the text was written, and how it was circulated in China, or more importantly, what the traditional Chinese views about war are. However, for target readers thou-sands of miles away from China and two thousand years later, the under-standing of Sun Tzu may be rather difficult as few of these readers have been exposed to such background knowledge. It is definitely a demanding and almost impossible task for ordinary target readers to detect and screen relevant information about Sun Tzu from voluminous Chinese literature elsewhere. In addition, some deficiencies and shortcomings that exist in the translation of the strategic culture in the core text also need the paratexts to offer some help. Therefore, to ensure a better reception *of AOW* among the Western audience, some additional information must be supplemented by the paratexts.

In TT1, Giles chooses to add into the paratexts the necessary information about Sun Tzu and his military classic, so as to make up for the limitations of the core text. Supplementation in TT1 for the ancient Chinese strategic culture falls into three major aspects: an introduction to the author Sun Tzu, interpretation and substantiation of Sun Tzu's strategic thinking and descriptions of the different versions and the impact of the book.

As we have pointed out, authorship of a classic is an important matter for both the SL and TL readers. Giles does not neglect this. Citing a large sum of references by famous Chinese scholars, he believes that Sun Tzu was a historical personage, and his book came into existence during the period 505–496 BC. The Biography of Sun Tzu from *Shi Ch'i* and The Preface by Ts'ao Ts'ao from the ST are translated in their entireties. He renders many paragraphs from "Sun Tzu Hsu Lu" into English to support his viewpoint. He also quotes from many resources that are not included in the ST. In addition to the authorship of the core text, Giles makes efforts to inform English readers of the authors of the Chinese comments. In "Commentators", a newly compiled paratext, Giles introduces the eleven Chinese scholars and generals who remarked on Sun Tzu's thirteen chapters. It reports their life stories, their military feats and the features of their comments.

The great majority of Giles' paratexts involve the explanation and substantiation of Sun Tzu's strategic thinking in the core text. Giles does not fully translate the 2,760 items of comments in 133,500 words from the ST. Instead, he selectively translates and sometimes rewrites them into about 550 notes in 46,200 words. He omits the repetitive parts, summarizes similar statements and chooses the comments he considers more important for English readers.

Many terms about military strategy and tactics, weapons, terrains and organizations are explained with more detail in the notes either translated from Chinese comments or authored by Giles himself. More often than not, when the military terms and principles are domesticated in the core text, the notes compensate for the loss of foreign flavor. When the military terms are foreignized in the core text, the notes provide further interpretation to ensure accessibility and readability. These notes are helpful in the adequate, though not complete, reconstruction of ancient Chinese military culture. For instance, ancient Chinese weapons and equipment, such as swift chariots (驰车), heavy chariots (革车) and large shields (橹), are further explained in TT1 notes.

Example 9 Dialectical view (p. 34–35)

ST: 三军之众，可使必受敌而无败者，奇正是也。

TT1: To ensure that your whole host may withstand the brunt of an enemy's attack and remain unshaken – this is effected by **manoeuvers direct and indirect**.

Note: (1) For 必，there is another reading 毕, "all together," adopted by Wang Hsi and Chang Yü. (2) We now come to one of the most interesting parts of Sun Tzu's treatise, the discussion of the 正 and the 奇. (3) As it is by no means easy to grasp the full signification of these two terms, or to render them at all consistently by good English equivalents, it may be as well to tabulate some of the commentators' remarks on the subject before proceeding further. (4) Li Ch'üan: "Facing the enemy is **chéng,** making literal diversion is **ch' i**" 当敌为正傍出为奇... (5) To put it perhaps a little more clearly: any attack or other operation is 正, on which the enemy has had his attention fixed; whereas that is 奇, which takes him by surprise or comes from an unexpected quarter. (6) If the enemy perceives a moment that is meant to be 奇，it immediately becomes 正.

[There are over 550 words in the note translated by Giles. Owing to limited space here, only Li Ch'üan's comment is recorded to give a glimpse of Giles's practice. The sentences are numbered by the author of this study.]

For important military concepts, Giles seldom hesitates to translate notes from ST comments. In Example 9, Giles paraphrases "奇正" into "manoeuvers direct and indirect" with a domesticating approach, sacrificing the Chinese cultural essence embodied in the phrase. In order to make up for such deficiency, Giles makes a long note of over 550 words to supply additional information. He admits firstly that the terms 奇 and 正 are significant and interesting but difficult to match with any English terms. He then translates six comments by Chinese scholars to illustrate the meaning of the paired concepts from different perspectives. At the end of the note, Giles offers his own summarizing remarks to further clarify them. All through the note, Giles carries the Chinese ideographic characters "奇" and "正" directly into the English sentence, or sometimes uses the Chinese pronunciation *ch' i* and *chéng*, to keep the original flavor. Such practice is a close collaboration between Giles' translation of the core text and the note: the former adopts a domestication approach aiming to ensure the intelligibility of the concepts while the latter uses foreignization providing a chance for the translation to retain the exotic cultural features.

As we have mentioned above, Sun Tzu's argumentation in the SL core text is rather aphoristic and laconic, usually lacking concrete supportive facts. Fortunately, such deficiency is remedied by the SL paratexts with rich examples of many reputed battles, generals and politicians in Chinese history. Correspondingly, in the paratexts of TT1, examples such as the battle by Fei

River, the campaign commanded by Ts'ao Gui as well as the rivalry between Sun Pin and Pang Chuan are translated selectively by Giles to substantiate Sun Tzu's strategic principles in the translated core text.

Example 10 Foreknowledge and information

ST: 故曰：知己知彼，百战不殆；不知彼而知己，一胜一负；不知彼不知己，每战必殆。

TT1: Hence the saying: If you know the enemy and know yourself, you need not fear the result of a hundred battles. If you know yourself but not the enemy, for every victory gained you will also suffer a defeat. If you know neither the enemy nor yourself, you will succumb in every battle.

Note: Li Quan cites the case of Fu Jian 苻坚，prince of Qin 秦，who in 383 A.D. marched with a vast army against the Jin 晋 Emperor. When warned not to despise an enemy who could command the services of such men as Xie An 谢安 and Huan Chong 桓冲，he boastfully replied: "I have the population of eight provinces at my back, infantry and horsemen to the number of one million; why, they could dam up the Yangtsze River itself by merely throwing their whips into the stream. What danger have I to fear?" Nevertheless, his forces were soon after disastrously routed at the Fei Shui River, and he was obliged to beat a hasty retreat.

Example 10 sets forth Sun Tzu's most reputed principle of foreknowledge and information. The translation closely follows the pattern of ST, rendering the aphorism into a parallelism. However, there is no evidence or exemplification provided in both source core text and target core text. However, the note translated from Chinese comments provides a vivid example, the battle by the Feishui River, a famous military event in ancient China, during which one party ended up being defeated because they were arrogant and ignorant of their adversary. This exemplifying paratext becomes a strong support to Sun Tzu's strategic principle of knowing the enemy and knowing yourself, helping target readers better understand Sun Tzu's strategic ideas.

It should be noted that, some paratexts in TT1 aim to highlight Sun Tzu's notion of peace and the opposition to militarism, which is the core value of Chinese strategic culture.

In "Apologies for War", a paratext in TT1, Giles (1910, xliii) remarks that China is "the greatest peace-loving nation on earth". In spite of the experience in the turbulence of wars, "the great body of Chinese sentiment… has been consistently pacific and intensely opposed to militarism in any form" (Giles 1910, xliv). In addition, Giles (1910, xlv) cites the great historian Ssuma Ch'ien, who says: "[m]ilitary weapons are the means used by the Sage to punish violence and cruelty, to give peace to troublous times, to remove difficulties and dangers, and to succour those who are in peril". Tu Mu, a great scholar in Tang Dynasty, is also quoted: "war may be defined as publishment, which is one of the functions of government" (Giles 1910, xlv) and "the end in

view is to get rid of wicked people, and to give comfort and relief to the good"
(Giles 1910, xlvi). It is also stated that:

> Weapons are baneful and fighting perilous; and unless a general is in con-
> stant practice, he ought not to hazard other men's lives in battle. Hence it
> is essential that Sun Tzu's 13 chapters should be studied…The treacherous
> and underhand nature of war necessitates the use of guile and stratagem
> suited to the occasion. There is a case on record of Confucius himself
> having violated an extorted oath, and also of his having left the Song
> State in disguise. Can we then recklessly arraign Sun Tzu for disregarding
> truth and honesty?
>
> (Giles 1910, 45–46)

In addition to the introductory paratext, Giles also rewrites notes in order to
highlight Sun Tzu's peace-orientation. The following example is a case in point.

Example 11 (p. 162)

ST: 非人之将也，非主之佐也，非胜之主也。

TT1: One who acts thus is no leader of men, no present help to his sover-
eign, no mastery of the country.

Note: This idea, that the true object of war is peace, has its root in the
national temperament of the Chinese. Even so far back as 597 B.C., these
memorable words were uttered by Prince 庄 Chuang of the Ch'u State: 夫
文止戈为武…夫武禁暴戢兵保大定功安民和众丰财者也 "The character
for 'prowess' (武）is made up of 'to stay' 止 and 'a spear' 戈 (cessation
of hostilities)." Military prowess is seen in the repression of cruelty, the
calling in of weapons, the preservation of the appointment of Heaven,
the firm establishment of merit, the bestowal of happiness on the people,
putting harmony between the princes, the diffusion of wealth. [Tsochuan
宣公XII. 3 ad fin]

Example 11 is from the 13th chapter "The Use of Spies". In ST, Sun Tzu
stresses on the importance of espionage by saying to the effect that one cannot
be the leader of the army, the assistant to his state, and the expert of victory
if he remains ignorant of the enemy's condition simply because he begrudges
the expense of the spies. The note(s) for this section contain ten comments by
Chinese scholars to explain the Chinese words such as 人 and 佐. Some reiterated the importance of espionage.

The core text is a rather faithfully translated. However, Giles does not
translate the original ST note attached to this sentence, instead, he replaces
it with a new note. The new note begins by saying that, for Chinese people,
"the true object of war is peace". It then explains Chinese peace-orientation
with an analysis of the Chinese character 武. With this note, Giles highlights the notion of peace in Chinese strategic culture. Although it might be

inappropriate for Giles to add such a note to a sentence discussing spies, Giles' efforts are excusable and even praiseworthy because they reveal the core value of Chinese strategic culture which stresses on peace, distinguishing itself from Western military culture.

These paratexts added by Giles expresses a deep understanding of peace-orientation: the most important feature of Sun Tzu's strategic thinking. Giles also reckons that the excuse for deception in war comes from peace orientation. Deception, which is considered by Confucians as a demerit disregarding the virtue of truth and honesty, is justified of its application in warfare. With the manipulation on the content of paratexts, Giles tries to impress the readers with Sun Tzu's peace-orientation in strategy.

Giles also takes pains to trace the origin of *AOW* and its different versions throughout generations, as well as its influence upon Chinese generals. He highlights three editions that may stay closest to the original: one existing in T'ung Tien (通典) in the middle of the Tang Dynasty (618–907), another edited by Chi Tien-pao (吉天宝) during the Sung Dynasty (960–1279) and the third edited by Sun Hsing-yen (孙星衍) and Wu Jen-chi (吴人骥) in Qing Dynasty (1636–1912). The Sun Hsing-yen and Wu Jen-chi version is commonly denominated as the standard text of Sun Tzu due to the editors' careful comparison and correction. Giles also explains why Sun Hsing-yen and Wu Jen-chi version is chosen as the ST. In the paratext "Appreciations of Sun Tzu", the profound impact of Sun Tzu's text upon famous generals such as Han Hsin (韩信) and Yo Fei (岳飞), as well as scholars such as Su Hsün (苏洵) and Cheng Hou (郑厚) is stated.

The above analysis reveals that supplementation in Giles' paratexts is conducive to the reconstruction of ancient Chinese strategic culture as it not only interprets and substantiates Sun Tzu's strategic principles but also highlights the peace-orientation of Sun Tzu.

Comparison

In addition to supplementation, Giles adopts another approach in his paratexts: to juxtapose and compare Chinese strategic culture with Western counterparts. Among the large mass of notes, more than 50 refer to Western military tradition in a wide range of topics: military concepts, books, generals and battles. Their common purpose is to illustrate, support and confirm "the points in old Sun Tzu" (Ball 1910, 964).

In these notes, 22 famous Western generals are referred to and some repeatedly. For instance, Giles quotes Napoleon and Henderson eight times, Hannibal five, the Duke of Wellington four and Maréchal Turenne, Frederick the Great and Julius Caesar three. Over 13 famous battles led by Western generals are listed to prove the truth of Sun Tzu's military theory, such as the battles of Waterloo (Giles 1910, 5, 48, 130), Trafalgar (Giles 1910, 57), and Marengo (Giles 1910, 57). Some Western military concepts such as Fabian tactics (Giles 1910, 50) and theories such as Frederick the Great's classification of spies (Giles 1910, 168) are also mentioned.

Giles also cites 13 Western military books, among which *Aids to Scouting* is quoted four times and *The Science of War* twice. *Aids to Scouting*, a handbook published in 1899 as a title of Gale and Polden's Military Series, was written by General Robert Baden-Powell on scouting training for the military. It includes details of many subjects such as tracking, noticing signs and deducing their meanings and keeping hidden. As pointed out by Minford (2008, xxii), "Giles is constantly looking for contemporary resonances when he sees a link between Sun Tzu's thinking and the development of 'scouting' as a branch of army training".

Example 12 (p. 17)

ST：是故百战百胜，非善之善者也。不战而屈人之兵，善之善者也。

TT1：Hence to fight and conquer in all your battles is not supreme excellence; supreme excellence consists in breaking the enemy's resistance without fighting.

Note: Here again, no modern strategist will but approve the words of the old Chinese general. Moltke's greatest triumph, the capitulation of the huge French army at Sedan, was won practically without bloodshed.

This example declares the principle of "victory without fighting", as the cornerstone for Sun Tzu's military theory. It calls for the end of military conflict by diplomatic, economic or other less destructive means to secure the maximum success with minimum loss. In the core text, Giles interprets Sun Tzu's message rather clearly; while in the note, he cites the well-known Battle of Sedan led by Helmuth von Moltke (1800–1891) to endorse Sun Tzu's philosophy. Helmuth von Moltke was a German Field Marshal and the chief of staff of the Prussian Army for thirty years, famed for many successful military operations. In September 1870, he commanded at the Battle of Sedan fought between France and Prussia. On September 1, the Prussian soldiers managed to encircle the French army at the Sedan Fortress and effectively defeated the French resistance. The French army suffered heavy losses but found no way to escape. The next day, the French surrendered, which resulted in the capture of the French Emperor Napoleon III and his 80,000 men (Moltke 1893, 99–100; Wawro 2005, 228). By means of siege, the Prussians avoided heavy casualties on their own side, inflicted heavy casualties on the French side and won a great victory. Therefore, the victory at Sedan testifies to what Sun Tzu preached 2,000 years ago.

Most of the quotes from Western strategic culture, like the one in Example 12, provide chances for comparison. The comparison shows that the military philosophy of a single ancient Chinese strategist is shared and supported by 22 Western generals, both ancient and modern. In this way, Sun Tzu's teachings are proved applicable and timeless throughout generations and across national boundaries.

Although most of the 50 notes of comparison stress on the similarity between the Chinese and Western strategic culture, several point out the differences between them. For instance, Giles (1910, 45) states in one note that "Sun Tzu, unlike certain generals in the late Boer war, was no believer in frontal attacks". In another note, after having compared Sun Tzu's classification of spies and that of Frederick the Great, Giles (1910, 168) concludes that Frederick's classification is "a bad cross-division". These differences evidently highlight the uniqueness and worth of Sun Tzu's teaching to Western strategists.

Evaluation

In addition to supplementing information and comparing Chinese and Western strategic culture, Giles never hesitates in the paratexts to express his attitude toward and judgment on Sun Tzu. Paratexts, such as the Preface, Introduction and notes, reveal that Giles thinks critically about Sun Tzu's military thesis and that Giles' evaluation toward Chinese strategic culture is positive in general.

In the Preface, Giles (1910, vii–viii) sets a tone of his attitude toward *AOW*: "highly valued in China as by far the oldest and best compendium of military science". The word "oldest" brings out the sense of the long-standing tradition of the book, "best" stress on its high quality and the phrase "highly valued" expresses the degree of reception among the SL audience.

In the Introduction, specifically "Sun Tzu and His Book", Giles pays tribute to Sun Tzu and his military expertise:

> [T]heir essence has been distilled from a large store of personal observation and experience. They reflect the mind not only of a born strategist, gifted with a rare faculty of generalisation, but also of a practical soldier closely acquainted with the military conditions of his time. To say nothing of the fact that these sayings have been accepted and endorsed by all the greatest captains of Chinese history, they offer a combination of freshness and sincerity, acuteness and common sense, which quite excludes the idea that they were artificially concocted in the study.
>
> (Giles 1910, xxv–xxvi)

In this paragraph, Giles eulogizes Sun Tzu as "a born strategist" and "a practical soldier" who knows how to generalize general principles and adjust to his current military context. Sun Tzu's strategic thoughts are exalted as "a combination of freshness and sincerity, acuteness and common sense", which suggests the preeminent qualities of *AOW* in different aspects. These evaluative remarks are helpful in reconstructing a positive image of Sun Tzu.

More than 20 praising remarks are scattered among Giles' notes to demonstrate that he believes the teachings in *AOW* are true, effective and respectable.

Here are just a few examples: "Another sound piece of military theory" (Giles 1910, 18); "Here begin Sun Tzu's remarks on the reading of signs, much of which is so good" (Giles 1910, 88); "Those who may think that Sun Tzu is over-emphatic on this point would do well to read Col. Henderson's remarks" (Giles 1910, 131); and "Sun Tzu's argument is certainly ingenious" (Giles 1910, 163). For more details, let us see the following example.

Example 13 (p. 6)

ST: 兵者诡道也。

TT1: All warfare is based on deception.

Note: The truth of this pithy and profound saying will be admitted by every soldier. Col. Henderson tells us that Wellington, great in so many military qualities, was especially distinguished by "the extraordinary skill with which he concealed his movements and deceived both friend and foe."

Deception (诡道) is a fundamental principle of Sun Tzu's military theory. In Example 13, this principle has been clearly and gracefully translated in the core text. In the note, Giles provides an acclamatory remark. By using the words "truth" and "profound", he states that Sun Tzu's thought-provoking principle is true and applicable in military conflicts. By using the phrase "every soldier", Giles stresses its universal acceptance among military men, ancient or modern, East or West, young or old. This evaluative remark, short yet powerful, clearly conveys Giles' positive attitude toward target readers. Such praise is helpful in developing readers' respect toward Sun Tzu's military ideas. In addition, Giles' quotation of Henderson's comment on Wellington's outstanding military quality also exemplifies Sun Tzu's military wisdom.

Giles also criticizes Sun Tzu in four notes. Three out of the four criticizing notes are against Sun Tzu's way of presenting the text. For example, Giles is not satisfied with the way Sun Tzu used imperatives in some clauses (Giles 1910, 43). Only one criticizing note focuses on Sun Tzu's military theory. In the beginning of the tenth Chapter, Sun Tzu distinguished six kinds of terrains: accessible ground, entangling ground, termporizing ground, narrow passes, precipitous heights, and positions at a great distance from the enemy. Giles comments that there is "a faultiness of this classification. A strange lack of logical perception is shown in the Chinaman's unquestioning acceptance of glaring cross-divisions such as the above" (Giles 1910, 100).

Overall Giles' notes of evaluation show his own positive attitude toward Chinese military culture, and are helpful in fostering among Western readers the appreciation and respect toward Sun Tzu's military thesis, which is quite important in the reception of a translation.

To sum up, our analysis of Giles' paratexts discovers that the ancient Chinese military culture is adequately supplemented, compared and positively evaluated in the paratexts including the translated Chinese comments, newly written preface, introduction and added notes. These approaches of supplementation, comparison and evaluation constitute a necessary and conducive aid to reconstructing ancient Chinese military culture. Since Giles makes little effort to put *AOW* into his contemporary setting and systematically interpret them in modern terms, we tend to assume that he seldom adopts the approach of recontextualization in the paratexts.

4.3 Strategic culture reconstruction in Griffith's translation

In this section, we examine the way in which the core text is translated and the methods with which its paratexts are dealt with, to find out how the strategic culture is reconstructed in Griffith's translation (TT2).

4.3.1 Translation of the core text

The examination on the translation of the core text covers three aspects: what approaches and methods are employed in translating military terms; how the strategic principles are translated, and what happens to the stylistic features. For comparative purpose, examples of Chinese military terms, principles and metaphoric expressions used in analysis of TT2 are in most cases the same as the ones used in analysis of TT1.

Among the 385 sentences from the SL core text, only five are not translated into TT2. These omitted ones are just a few and they do not include any of Sun Tzu's key concepts; therefore, it can be said that such omission is minute, leading to rather small damage to the overall meaning of the core text. As a result, TT2 is almost a full translation.

Translation of military terms

Our survey shows that in TT2, about 160 military terms (79% of the total) are translated with the domesticating approach, while 43 terms (21% of the total) are translated with the foreignization approach (see Table 4.7). It shares a similar trend with TT1 in the translation of military terms. The survey shows that the domestication approach far outweighs the foreignization approach. Like TT1, TT2 attaches greater importance to the readability of the text and acceptability of Sun Tzu's military terms. However, as indicated by our analysis, TT2 chooses to maintain the concise style of ST when translating the terms, which suggests that a large number of terms in TT2 are still in need of further interpretation.

Paraphrasing is the most frequently used method of translation, with about 70% of the military terms interpreted into culture-neutral words and phrases. For instance, the terms 夺气 and 夺心 are explained as "be robbed of its spirit" and "deprived of his courage". They are easily intelligible, however, the exotic Chinese cultural subtlety and profundity carried by "心" and "气" are lost.

Table 4.7 Translation approaches and methods for military terms in TT2

	Foreignization		Domestication			
	Calque	Transliteration	Paraphrasing	Generalization	Omission	Substitution
Strategy and tactics	16	1	102	1	2	2
Supply and armament	4	0	17	5	0	2
Terrain	11	0	0	0	0	0
Organization	3	0	0	1	1	3
Battles and officers	1	1	13	0	0	0
Total (100%)	35 (18%)	2 (1%)	132 (71%)	7 (4%)	3 (2%)	7 (4%)

Substitution, another domesticating method, is used to translate seven military terms, but at the cost of Chinese cultural essence. For instance, in ancient China, 钟 means a unit of measurement equal to the volume of eight cauldrons. However, it is translated as "bushel", a Western measurement equal to 35.42 liters. Another example, 廊庙, meaning the great temple in which the Chinese king discusses national affairs with his ministers, is rendered as "temple council". This translation is problematic to some extent since the council in Western political systems is quite different from a discussion in Chinese temple, and the strange juxtaposition of "temple" and "council" seems contradictory. In this translation, the exotic flavor is lost in some degree and the cultural difference between East and West is removed if no additional information is added.

The generalization method is also used to translate seven terms. For instance, 辎辐 is generalized as "the necessary arms and equipment". Such translation is easy to understand, but it conceals the basic difference between 辎辐 and other military equipment and fails to bring out its distinct feature.

Only a small number of military terms in the core text are translated with the foreignizing approach and at the same time, TT2 manages to maintain the concise style of the ST. This approach is employed to such an effect that the Chinese cultural essence is mostly maintained in the core text, but a need for further interpretation surfaces.

Calque is a method used over 30 times in TT2 to foreignize the military terms. For example, 轻车 is calqued as "light chariots". The translation is a short phrase indicating a vehicle, but there is no adequate information to illustrate its usage as the mainstay of the assaulting troops with high speed and great mobility in ancient Chinese battlefield. 庙算, is rendered word-by-word as "estimates made in the temple". Abbreviations such as 五行 are translated into "five elements", 黄帝 "the Yellow Emperor", and 四帝 "four sovereigns". However, in the core text, there is no interpretation to these terms, which fails to reveal the cultural connotation embodied within.

Transliteration is another foreignizing method employed in TT2. The names of the ancient strategists and kings are in most cases transliterated. For example, 伊挚 is transliterated as "I Chih" in TT2; 吕牙 as Lu Yu; 刿 as "Tso Kuei"; 诸 as "Chuan Chu". Although TT2 has provided both surnames and given names of these strategists, little information about their military talent is given in the core text. These names may become obscure entities for readers if no additional information is added.

The analysis of the translation of military terms in TT2 reveals that although most terms are translated in a fluent, transparent way with the domesticating approach, extra help is still needed to bring out their cultural essence to the fullest extent.

Translation of strategic principles

Next, we will look at the key principles of Sun Tzu's strategic theory and their translation in TT2. Sentences that best illustrate these five principles are also selected as examples.

Example 14 Victory without fighting (chapter 3)

ST: See Example 2

TT2: (1) Generally in war the best policy is to take a state intact; to ruin it is inferior to this. To capture the enemy's army is better than to destroy it; to take intact a battalion, a company or a five-man squad is better than to destroy them. (2) For to win one hundred victories in one hundred battles is not the acme of skill. To subdue the enemy without fighting is the acme of skill. (3) Thus, what is of supreme importance in war is to attack the enemy's strategy. Next best is to disrupt his alliances. The next best is to attack his army. The worst policy is to attack cities. (4) Attack cities only when there is no alternative. [The numbers are added by the present author]

This example explains Sun Tzu's key strategic principles. In the first sentence of ST, each of the four terms for military organization in ancient China, 军, 旅, 卒 and 伍, refers to a military unit with about 10,000 soldiers, 500, 100 and 5 respectively. These terms are translated into "an army", "a battalion", "a company" and "a five-man squad". In the west, an army usually holds 100,000 soldiers; a battalion 400 to 1000 soldiers; a company, 100 to 250; a squad, 8 to 25 soldiers. Compared with TT1, the wording in TT2 is closer to the ST in reflecting numerical scales of the four units.

The phrase 百战百胜 in the second sentence is literally translated into "win one hundred victories in one hundred battles" in TT2. 百 in Chinese can either refer to an accurate number of "one hundred", or simply suggest "a large number". In this phrase, it is used rhetorically to convey the meaning of the latter. The literal translation in TT2 retains the flavor of Chinese expression, but Western readers may mistake it for an exact number if no explanation is added.

The third sentence lists four forms of warfare in a hierarchic order from the best to the worst: the game of strategy, the wrestle of diplomatic efforts, the battle in field and the attack against the walled cities. In TT2, 伐谋 is rendered into "to attack the enemy's strategy", and 伐交 into "to disrupt his alliances", representing the first two of the four scales of warfare proposed by Sun Tzu. With this faithful translation of the ST, TT2 highlights the non-military measures and articulates Sun Tzu's principle of victory without fighting. As a result, TT2 has done a better job than TT1 in translating Sun Tzu's first principle.

Example 15 Holistic approach (chapter 2)

ST: See Example 3

TT2: (1) Victory is the main object in war. If this is long delayed, weapons are blunted and morale depressed. When troops attack cities, their strength will be exhausted. (2) When the army engages in protracted campaigns the resources of the state will not suffice…(3) Thus, while we have heard of blundering swiftness in war, we have not yet seen a clever operation that was prolonged. (4) For there has never been a protracted war from which a country has benefited.

Example 15 has illustrated Sun Tzu's holistic approach toward war, which takes into consideration military affairs, economic issues and psychological factors. Generally speaking, TT2 faithfully translates sentences two and four. 钝兵 is translated into "weapons are blunted" indicating physical strength, 挫锐 "morale depressed" suggesting the psychological factor, 国用 "the resources of the state" as well as 不足 "not suffice" meaning economic issues. In these well-structured sentences, the translated terms express neatly Sun Tzu's overall approach to the warfare.

However, TT2 translates 拙速 in sentence three as "blundering swiftness", which is similar to the TT1 version "stupid haste". With this word-for-word translation, there emerges a contradiction: Sun Tzu thinks that swiftness is blundering while he also encourages the quick ending of war. At the same time, the core text adds no extra information to explain "blundering swiftness", which can easily lead to confusion among readers. Therefore, such contradiction needs to be redressed by other means.

Example 16 Foreknowledge and information chapter 3)

ST: See Example 4

TT2: Therefore I say: Know the enemy and know yourself; in a hundred battles you will never be in peril. When you are ignorant of the enemy but know yourself, your chances of winning or losing are equal. If ignorant both of your enemy and of yourself, you are certain in every battle to be in peril.

In Example 16, Sun Tzu puts forward a universal law for victory in war: adequate foreknowledge of the enemy and the self. In TT2, the parallel structure is not strictly followed, but the wording is rather concise and graceful. TT2 consistently uses "know" for 知, "ignorant" for 不知 and "in peril" for 殆. It paraphrases 百战不殆 into "never be in peril", 一胜一负 into "chances of winning or losing are equal". Generally speaking, TT2 has offered a clear and acceptable translation of Sun Tzu's strategic principle about foreknowledge, but fails to retain the rhetorical effect of ST.

Example 17 Dialectical view (chapter 5)

ST: See Example 5

TT2: (1) That the army is certain to sustain the enemy's attack without suffering defeat is due to operations of the **extraordinary** and the **normal forces**. (2) Troops thrown against the enemy as a grindstone against eggs is an example of **a solid acting upon a void**. (3) Generally, in battle, use the normal force to engage; use the extraordinary to win. (4) Now the resources of those skilled in the use of extraordinary forces are as infinite as the heavens and earth; as inexhaustible as the flow of the great rivers.

Example 17 expounds on two pairs of concepts, 奇 and 正, 虚 and 实, which reflect the dialectical view on military tactics held by Sun Tzu. Like TT1, TT2 opts for a domesticated translation. In TT2, 正 is explained as "the normal forces", and 奇 as "extraordinary forces". Another pair of concepts, 虚 and 实, are interpreted as "a void" and "a solid". However, these translations have deprived many alternative interpretations of these concepts. Like in ST, no definitions of these concepts or illustrations of the principles are furnished in the core text. The interpretation method in the core text, in spite of its intelligibility and conciseness, has its limit in conveying the multiple and profound meanings of military terms and consequently the principles themselves.

Example 18 Deception, maneuver and flexibility (chapter 1)

ST: See Example 6

TT2: (1) All warfare is based on deception. (2) Therefore, when capable, feign incapacity; when active, inactivity. (3) When near, make it appear that you are far away; when far away, that you are near. (4) Offer the enemy a bait to lure him; feign disorder and strike him. (5) When he concentrates, prepare against him; where he is strong, avoid him. (6) Anger his general and confuse him. Pretend inferiority and encourage his arrogance. (7) Keep him under a strain and wear him down. When he is united, divide him. (8) Attack where he is unprepared; sally out when he does not expect you. (9) These are the strategist's keys to victory. It is not possible to discuss them beforehand.

In Example 18, Sun Tzu declares another key principle in strategy: 诡道. Like TT1, TT2 translates it as a derogatory term "deception", which slightly changes its shades of meaning from neutral to negative.

In the SL sentences from the second through to the eighth, subjects are omitted for the sake of brevity. This leads to multiple interpretations of these sentences if different subjects are added. Like TT1, there is only one explanation noted down for each sentence in TT2, which reduces the multiplicity of the ST interpretation.

This example also shows that ST is good at noting down the principle of deception and the guidelines for its application, but it does not provide any battle example to illustrate and substantiate the principle and rules. TT2 follows ST, which suggests that extra information is needed elsewhere to help Western readers fully understand them.

The above analysis of translation of the five principles shows that Sun Tzu's strategic ideas are on the whole faithfully conveyed. Most of the military terms are paraphrased and the concise style of ST is followed. However, the exotic flavor of them is somewhat lost. The multiple and profound meanings of some military terms and principles are not carried over to the fullest extent. The deficiency of the core text in definition, illustration and substantiation for military concepts is closely inherited in TT2.

Translation of metaphoric expressions

As we have mentioned in Section 4.1, metaphoric expressions are essential in the representation of Sun Tzu's strategic ideas. According to our survey, 25 Chinese metaphors and similes (89% of the total) are preserved in TT2 while only two are paraphrased (See Table 4.8), which means foreignizing approach is more frequently used than the domesticating approach.

Example 19 Simile

ST: See Example 7

TT2: (1) Now an army may be likened to water, for just as flowing water avoids the heights and hastens to the lowlands, so an army avoids strength and strikes weakness. (2) And as water shapes its flow in accordance with the ground, so an army manages its victory in accordance with the situation of the enemy. (3) And as water has no constant form, there are in war no constant conditions.

Table 4.8 Translation of metaphors and similes in TT2

	Total	Foreignization	Domestication
Simile	21	20 (95%)	1 (5%)
Metaphor	6	5 (83%)	1 (17%)
Overall	27	25 (92%)	2 (8%)

In Example 19, the maneuvering of an army is vividly compared to the flowing movement of water. TT2 adopts a foreignizing approach, clings to the rhetoric device and inform the readers the Chinese way of military thinking. As water flowing is a universal phenomenon, it is easy for the readers to get the gist of such a simile. Thus, foreignizing translation in this sense is sufficient to maintain the mode of thinking and the rhetorical style of the core text.

Example 20 Metaphor

ST: See Example 8

TT2: The doctrine of war is to follow the enemy situation in order to decide on battle.

In Example 20, 践墨 means that an ancient Chinese carpenter follows the marks that are made by a thread saturated with ink while he is cutting through the wood. The metaphor 践墨 is paraphrased as "follow the enemy situation", an expression similar to "walk in the path defined by rule" in TT1. The domesticating translation roughly expresses the meaning of the metaphor, but it sacrifices the exotic flavor and vividness, most likely for the sake of economy in the core text.

The analysis in this section reveals that in translating military terms in the core text, domestication is the most frequently used approach, and paraphrasing is the most often used method in Griffith's translation. TT2 stresses on intelligibility and acceptability, at the same time, it chooses to retain the concise and vivid style of ST. Compared with TT1, TT2 comes up with a more faithful translation of the military terms and strategic principles, but not without shortcomings. The profundity and exotic flavor of Chinese strategic culture is somewhat sacrificed in this translation of the core text.

4.3.2 Role of paratexts

Rather different from TT1, whose paratexts translated from ST far outnumber the added ones, in TT2, the paratexts translated from ST are reduced to about 13,200 words, while the added ones amount to 41,100 words (see Table 4.9). This means TT2 attaches greater importance to newly added paratexts.

There are altogether seven major paratexts. Among all the paratexts, comments translated from the ST are placed very close to the core text, right after each English sentence. They mainly explain the meaning and support the argument of Sun Tzu. This practice is very similar to that of TT1.

"Translator's Notes" consist of three major parts: further interpretation to the core text, explanation to translator's choice and occasional reference to Giles' translation. More details of the notes will be given in the following part of this section.

"Biography of Sun Tzu" is a paratext fully translated from the Shih-chi, where Sun Tzu's legendary story of training the court concubines is vividly told. This is done in a way similar to TT1.

Table 4.9 Paratexts of TT2

	Title	Sub-titles	Relations with ST	Words
1	Chinese Commentary		Partially translated from ST comments	11,558
2	Translators' Notes		Composed by Griffith	5,411
3	Biography of Sun Tzu		Fully translated Biography of Sun Tzu from *Shi Chi*	1,710
4	Foreword		Written by Liddell Hart	1,004
5	Preface		Written by Samuel Griffith	940
6	Introduction (by Griffith)	(1) The author	Added	3,615
		(2) The text	Added	2,055
		(3) The warring states	Added	3,360
		(4) War in Sun Tzu's age	Added	2,908
		(5) Sun Tzu on War	Added	1,951
		(6) Sun Tzu and Mao Tse-tung	Added	3,806
7	Appendix (by Griffith)	(1) Wu Ch'i's Art of War	Added	7,956
		(2) Sun Tzu's Influence on Japanese Military Thought	Added	4,174
		(3) Sun Tzu in Western Languages	Added	1,491
		(4) Brief Biographies of the Commentators	Added	1,106
		(5) Bibliography	Added	1,345
Total				54,390

"Foreword" is a piece authored by Liddell Hart, the famous British strategist. With profound insight and broad vision in strategy, Hart compares *AOW* with Clausewitz's *On War* and provides a favorable evaluation on Sun Tzu. This paratext is rather important in the dissemination of TT2 thanks to the prestigious status and fame of Hart in the science of strategy around the globe.

In "Preface", Griffith summarizes the principles of *AOW*, gives a very brief historical account of *AOW* translations, and explains his motive for translation. Griffith acknowledges *AOW*'s canonical position in Chinese military literature. Griffith complains that "none of the five English translations is satisfactory; even that of Lionel Giles (1910) leaves much to be desired". He also reminds us that *AOW* is required reading for people who hope to gain better understanding of the grand strategy of China and Russia.

"Introduction" is a rather comprehensive and voluminous paratext with six subsections. (1) "The Author" gives a detailed introduction to the textual features of *AOW* to identify the time period in which Sun Tzu lived and wrote the book. (2) "The Text" is an expository essay that gives a narration of different versions of Sun Tzu's text. (3) "The Warring States" offers a panoramic view about the social and cultural context including philosophical thoughts, the political system, the economic situations, the use of iron and the need of strategists during this period. (4) "War in Sun Tzu's Age" focuses on the military context before and during Sun Tzu's time to explain the birth of his strategic ideas. (5) "Sun Tzu on War" again summarizes the principles in *AOW*, but in more detail this time. It highlights the principle of victory without fighting. It also points out the moral strength and intellectual faculty required of military leaders and the economic implications of war. The concepts of *shih* and *hsing* and *cheng* and *ch'i* are also discussed. The principles of deception, adaptability, foreknowledge and information are introduced. (6) "Sun Tzu and Mao Tse-tung" is an essay revealing the huge impact of Sun Tzu's upon Mao Tse-tung, a brilliant Chinese contemporary strategist. Sun Tzu's key principles are measured against several wars fought by the Red Army led by Mao Tse-tung.

TT2 also contains five appendixes. (1) *"Wu Ch'i's Art of War"* Is a translation of another famous strategic book written by Wu Ch'i who also lived in ancient China. (2) "Sun Tzu's Influence on Japanese Military Thought" This essay chronicles the spreading of *AOW* to Japan and its influence on Japanese warriors, strategists and generals in wars, from 760 AD to the end of the World War II. (3) "Sun Tzu in Western Languages" Introduces the translation of *AOW* into major Western languages: French, English, Russian and German. (4) "Brief Biographies of the Commentators" This piece lists the biographical notes of the Chinese generals and scholars who authored the comments in ST. (5) "Bibliography" Provides the references written in a variety of languages, Chinese, English and other Western languages.

The paratexts in TT2 cover a wider range of topics when compared with those of TT1. The main functions of this huge body of paratexts fall into four aspects: supplementation, recontextualization, comparison and evaluation.

Supplementation

Supplementation for Sun Tzu's strategic thinking covers four aspects: the interpretation as well as substantiation of Sun Tzu's strategic ideas, the introduction to the authorship, the different versions and the influence of *AOW*.

Interpretation of and substantiation to Sun Tzu's strategic thinking in the core text is usually realized through the Chinese comments and the translator's notes. Griffith selectively translates 208 comments of 11,600 words from about 2,760 comments of 133,500 words in the ST. Like Giles, he cuts out the repetitive parts and chooses the comments that he thinks can best help readers to understand Sun Tzu. In addition, he adds 192 notes, about 5,500 English words, to further supplement the core text.

It is evident that these comments translated from the ST paratexts and notes authored by Griffith, working in partnership with the TT2 core text, are helpful in the reconstruction of ancient Chinese strategic culture. In TT2, when most of the terms of military strategy, tactics, weapons, terrain and organizations are domesticated in core text, their literal meanings and transliterations are usually offered in the comments or notes. Sometimes, when concise interpretation fails to convey the profundity of these terms, they require further explanation. Examples of this practice include terms such as 道，天，法，千里马，金，形，势，分数，奇，and 正.

Example 21 (chapter 1)

ST: See Example 9

Core Text: That the army is certain to sustain the enemy's attack without suffering defeat is due to operations of the extraordinary and the normal forces, Troops thrown against the enemy as a grindstone against eggs is an example of a solid acting upon a void. Generally, in battle, use the normal force to engage; use the extraordinary to win.

Comments: (1) Li Ch'uan The force which confronts the enemy is the normal; that which goes to his flanks the extraordinary. No commander of an army can wrest the advantage from the enemy without extraordinary forces. (2) Ho Yen-his I make the enemy conceive my normal force to be the extraordinary and my extraordinary to be the normal. Moreover, the normal may become the extraordinary and vice versa. (3) Ts'ao Ts'ao Use the most solid to attack the most empty.

Translator's notes: The concept expressed by *cheng* (正)，"normal" (or "direct") and *ch'i* (奇)，"extraordinary" (or "indirect") is of basic importance. The normal (*cheng*) force fixes or distracts the enemy; the extraordinary (*ch'i*) forces act when and where their blows are not anticipated. Should the enemy perceive and respond to a *ch'i* maneuver in such a manner as to neutralize it, the maneuver would automatically become *cheng.*

Example 21 shows that Griffith both uses translated comments and adds notes to compensate for the core text lacks. In the core text, the concept of 奇 and 正 is paraphrased as "the extraordinary and the normal forces", to ensure comprehensibility and maintain the conciseness of the core text. Comments by three Chinese scholars, Li Ch'uan, Ho Yen-his and Ts'ao Ts'ao are translated to explain the profound implications of these concepts. In the translator's note, transliterations (*chéng* and *ch'i*) for them are provided together with the Chinese characters (正, 奇), which compensates for the lack of the exotic flavor in the core text and provides an opportunity for readers to access the original Chinese.

As in the ST core texts, Sun Tzu's military terms and principles are in need of concrete examples, Griffith translated 24 famous battles from the ST comments, such as the battle by Fei River, the campaign commanded by Ts'ao Gui, as well as the rivalry between Sun Pin and Pang Chuan for this purpose.

Example 22

ST: 利而诱之，乱而取之。

Core text: Offer the enemy a bait to lure him; feign disorder and strike him.

Comments: Tu Mu: The Chao general Li Mu released herds of cattle with their shepherds; when the Hsiung Nu had advanced a short distance he feigned a retirement, leaving behind several thousand men as if abandoning them. When the Khan heard this news he was delighted, and at the head of a strong force marched to the place. Li Mu put most of his troops into formations on the right and left wings, made a homing attack, crushed the Huns and slaughtered over one hundred thousand of their horsemen.

Example 22 showcases the supplementary role of paratexts. The SL core text is rather short and concise, using eight words to declare Sun Tzu's advice for military tactics. The TT2 core text conveys in a fluent manner the meaning of the ST: rendering 利 as "a bait", 诱 as "lure", 乱 as "feign disorder", and 取 as "strike". The translation here also follows the concise style of the ST with only 13 English words. To clarify Sun Tzu's highly condensed teachings, Griffith selected and translated from the SL comments a famous battle in Chinese history. In the battle, General Li Mu defeated the Khan's troops by releasing herds of cattle as bait and feigning disorder in retreatment. With the battle as case in point, it becomes easier for target readers to gain a better understanding of Sun Tzu's tactics.

Authorship is a frequently mentioned topic in TT2 paratexts. Griffith plunges deep into the lexical features of *AOW,* such as 霸王 (Hegemonic ruler)，金 (metallic currency) and 带甲 (armoured troops) to excavate evidence for the dating of *AOW.* He concludes that Sun Tzu probably lived around 400–320 BC. "But the originality, the consistent style, and the thematic development suggest that 'The Thirteen Chapters' is not a compilation, but was written by a singularly imaginative individual who had considerable practical experience in war". The Biography of Sun Tzu from Shi Ch'i is also fully translated in the paratexts.

Griffith also adds brief biographic information on the Chinese commentators. He introduces seven Chinese scholars and generals who made remarks on Sun Tzu's 13 chapters. With the largest portion, special emphasis is given to Ts'ao Ts'ao, who was entitled as Martial King.

Like Giles, Griffith puts great efforts in tracing the different editions of *AOW,* its circulation throughout dynasties, and its influence upon generals in China and in other countries. In "The Text", Griffith dates various editions in different dynasties from Chin, to Han, Tang, Sung, Yuan and finally Qing. He also points out that Sun Hsing-yen's edition is the very text he chooses to translate.

Recontextualization

If some paratexts supplementing information aims at an ancient Sun Tzu, then others in TT2 are trying to recontextualize Sun Tzu. *AOW* was written 2,500 years ago in northern China, while Griffith's translation was finished in 1963 and published in Oxford and New York. The huge spatial and chronological gap needs to be bridged. Paratexts then manage to put the ancient text *AOW* into a contemporary setting and name some Chinese military concepts with Western strategic terms. By means of recontextualization in the paratexts, a connection between the East and the West, the past and the present, is established.

When describing the profound impact of Sun Tzu's text, Griffith chooses Mao Tse-tung, one of the contemporary Chinese military talents, as a typical example. Mao was a legendary figure for the Western audience who could lead a small army in defeating his adversaries when severely outnumbered in a series of battles. In a paratextual essay entitled "Sun Tzu and Mao Tse-tung", Griffith (1963, 45) traces the origin of Mao Tse-tung's military strategy from the 1920s to 1950s and comes up with the conclusion that "Mao Tse-Tung has been strongly influenced by Sun Tzu's thought". He also probes into a series of wars: the campaigns launched by nationalists' troops against the Red Army in 1930s, the war against Japanese invasion from 1937 to 1945, the civil war between the communist Mao Tse-tung and Kuomingtang Chiang Kaishek from 1945 to 1949, and finally the Korean War from 1950–1953 in which the ill-fed Chinese Liberation Army defeated the well-equipped United Nation forces. These modern wars being guided by an ancient strategic mind testifies to the eternal appeal of the military principles coming from antiquity.

Griffith not only introduces the appeal of Sun Tzu within Chinese territory, but also informs that Sun Tzu has reached other nations such as Japan, France, Russia, Germany, the UK and the US. In two appendixes "Sun Tzu's Influence on Japanese Military Thought" and "Sun Tzu in Western Languages", Griffith introduces the impact of Sun Tzu's military theory in Western countries and Japan. Thus, Sun Tzu becomes a globalized strategist and *AOW* is reframed into an enormously expanded space via Griffith's paratexts.

In Griffith's Introduction, contemporary military terms are used to describe the warfare in Sun Tzu's time. For instance, "general's staff", a modern military term, is used to cover specialists needed in ancient Chinese battlefields: weather forecasters, map makers, commissionary officers, experts on crossing river, etc. (Griffith, 1963, 35). "Commander-in-Chief" is used for generals who were conferred upon supreme authority outside the capital (Griffith, 1963, 36). When Griffith writes about Sun Tzu's concept of secret agents: he uses the term "Fifth columns" (Griffith 1963, 44), which is coated with a sense of modernity. Fifth column is a term coined in 1936 during the Spanish Civil War, referring to a group of people who undermine a nation or a besiege city from within, in favor of an enemy group. Griffith also labels Sun

Tzu as "the first proponent of psychological warfare" since Sun Tzu advocates for placing attention on the morale and the hearts of the captains and soldiers (Griffith, 1963, 54). More importantly, Griffith finds that "Sun Tzu appreciated the difference between what we today define as 'national strategy' and 'military strategy'" (Griffith, 1963, 40). National strategy is a term in the West that emerged in the First World War. After 1914, it was defined by The US Department of Defense as the "art and science of developing and using the political, economic, and psychological powers of a nation, together with its armed forces during peace and war, to secure national objectives" (Bartholomees 2010, 8–9). Griffith also points out that "'*The Art of War*' is thus required reading for those who hope to gain a further understanding of the grand strategy of these two countries today" (Griffith 1963, xi). Grand strategy, is a concept firstly defined and systemically discussed in 1954 by Liddell Hart (Hart 1954, 335–336), which incorporates "the economic resources and man-power of nations" and also "the moral resources" in order to sustain the fighting services and to foster the people's willing spirit. This concept was soon picked up by many scholars such as John Collins (1973). Other contemporary terms used include: "the indirect approach" and "shock and elite troops" (Griffith 1963, 8) to describe military affairs in Sun Tzu's time.

These modern phrases have brought a new perspective from which *AOW* can be examined. With these paratexts, Sun Tzu is reincarnated into an expanded space, the setting of modern times when the translation was published. Sun Tzu's strategic thinking transcends over antiquity and great distance to reach a modern Western audience.

Comparison

In TT2 paratexts, comparison is also done between Sun Tzu's strategic wisdom and Western military thinking. Unlike TT1, which primarily seeks similarity between Eastern and Western military thoughts, TT2 covers the resemblance but stresses more on the differences.

In the Foreword by Liddell Hart, there is an in-depth comparison between Sun Tzu's *AOW* and Clausewitz's *On War* from different aspects. *On War*, authored by Prussian general and military theorist Carl van Clausewitz, is a classical and monumental military monograph that "moulded European military thought in the era preceding the First World War" (Hart 1963, v). Hart writes in a comparative setting that:

> Among all the military thinkers of the past, only Clausewitz (Griffith 1963) is comparable, and even he is more 'dated' than Sun Tzu, and in part antiquated, although he was writing more than two thousand years later. Sun Tzu has clearer vision, more profound insight, and eternal freshness.
>
> (Hart 1963, v)

Here, Hart reveals the outstanding quality of *AOW* by comparing Sun Tzu with Clausewitz. The phrase "clearer vision" suggests that Sun Tzu is more farsighted and versed in the issue of warfare and his theory is presented in a more limpid and vivid manner. Hart believes that there is a contrast between the "clarity of Sun Tzu's thought" and "the obscurity of Clausewitz" (Hart 1963, vi). Another phrase "more profound insight" describes the depth of Sun Tzu's perception of the principles of war. In Hart's view, "in that one short book was embodied almost as much about the fundamentals of strategy and tactics as I had covered in more than twenty books" (Hart, 1963 vii). The phrase "eternal freshness" pronounces the relevance of Sun Tzu's military theory in all ages, in comparison with the dated, antiquated Clausewitz.

Hart also identifies the contrast between "Sun Tzu's realism and moderation form" and "Clausewitz's tendency to emphasize the logical ideal and 'the absolute'" (Hart 1963, v). Sun Tzu is realistic and moderate because he believes in the victory without fighting and views war as the last resort. However, for Clausewitz the destruction of the enemy's force by military means is the most effective. Following Clausewitz's absolutist tendency, his disciples believe in "the theory and practice of 'total war' beyond all bounds of sense" (Hart 1963, v), which looks on all military and civilian resources and infrastructure as legitimate military targets, and especially involves the use of weapons and tactics that result in significant military and civilian casualties. This type of war may become unrestricted in terms of the weapons used, the territory or combatants involved. In this sense, Hart addresses Clausewitz and his disciples as "the Clausewitz extremists" and Sun Tzu as "the Chinese sage" (Hart 1963, vi).

Griffith, in his paratexts, also puts Sun Tzu and Western strategic theorists in comparative positions. He writes about the style of *AOW*: "Few military writers, including those most esteemed in the West, have stated this proposition as clearly as did Sun Tzu some twenty-three hundred years ago" (Griffith 1963, 40). While introducing Sun Tzu's chapter on the use of spies, Griffith writes:

> Fifth columns were as common in ancient China as in the Greek world and Sun Tzu takes account of them. The West has had considerable experience of this technique in recent years and our efforts to combat it cannot be described as entirely successful. Possibly Tu Mu's analysis of the types of men most susceptible to subversion is still worthy of examination.
>
> (Griffith 1963, 44)

The comparison between the espionage practices in China and in the West shows that *AOW* has offered a fuller and more systematic explanation on how spies can be best employed, and on how subversion can be cautioned against.

To summarize, the above comparison between the *AOW* and Western military masterpieces such as *On War* reveals that Sun Tzu outwits Clausewitz, which helps establish the highly prestigious status of Sun Tzu and paves the way for a better understanding and hospitable acceptance of *AOW* among the Western audience.

Evaluation

Translator's evaluative remarks can hardly be found in the translated core text of a classic due to its prestigious status and the code of conduct that requires neutrality and the least interference. However, in paratexts, translators or authors of paratexts are allowed the space and liberty to express their opinions. In the paratexts of TT2, rich commendatory remarks on the content, the style and the application of *AOW* are helpful in reconstructing Chinese strategic culture.

Griffith, the translator, himself acclaims Sun Tzu whenever it is possible in the paratexts. He labels *AOW* as "a thoughtful and comprehensive work, distinguished by qualities of perception and imagination which have for centuries assured it a pre-eminent position in the canon of Chinese military literature" (Griffith 1963, ix). It is thoughtful because it harbors "the first known attempt to formulate a rational basis for the planning and conduct of military operations" (Griffith 1963, x) and at the same time "a realistic basis" (Griffith 1963, 40). It is comprehensive since Sun Tzu's purpose is "to develop a systematic treatise to guide rulers and generals in the intelligent prosecution of successful war" (Griffith 1963, ix) and his perception of "a remarkable acuity" demonstrates "mental, moral, physical, and circumstantial factors that operate in war" (Griffith 1963, ix).

More importantly, the famous British strategist Liddell Hart joins his hands in praising Sun Tzu. The paratext Foreword by Hart brims with extolling remarks. It begins with the following two sentences:

> Sun Tzu's essays on '*The Art of War*' form the earliest of known treatises on the subject, but have never been surpassed in comprehensiveness and depth of understanding. They might well be termed the concentrated essence of wisdom on the conduct of war."
>
> (Hart 1963, v)

In these sentences, Hart acclaims that, as the earliest military thesis yet known, *AOW* features its comprehensive coverage and profound theory. It is hailed as the "concentrated essence of wisdom on the conduct of war". Wisdom means a lot more than sheer knowledge and experience, suggesting the sensible and acute accumulated knowledge and erudition. By using the word "wisdom", Hart highly underscores the intrinsic quality of Sun Tzu's strategic thinking. Furthermore, Hart uses the word "concentrated" to reveal the intensity of wisdom in the book and its concise linguistic style at the same time. Other phrases such as "clearer vision, more profound insight" (Hart 1963, v) are also used to acclaim for Sun Tzu's strategic depth and width.

In addition, Hart applauds the modern applicability of *AOW* by articulating phrases such as "eternal freshness" (Hart 1963, v) and "the agelessness of the more fundamental military ideas" (Hart 1963, vii). The evaluation on the modern relevance reveals the significance of the translation of the book for target readers.

With all the positive remarks, Hart crowns *AOW* in this way: "[i]n brief, Sun Tzu was the best short introduction to the study of warfare, and no less valuable for constant reference in extending study of the subject" (Hart 1963, vii). With the superlative phrase "the best", Hart manages to leave the readers an image of the most brilliant military thesis, which is remarkably helpful in the reception of the text.

4.4 Summary

In the textual analysis of ST, our investigation finds out that both the core text and its paratexts contribute to the canonical position of Sun Tzu in China. The core text, featuring profundity, conciseness and vividness, embodies Sun Tzu's military terms and strategic principles. The SL paratexts in large quantity, by supplementing and evaluating Sun Tzu's strategic thinking, constitute an important part of ancient Chinese military culture.

The two translations resort to a package of approaches and methods in both core text and paratexts to reconstruct the canonical military text. Generally speaking, in both TT1 and TT2, the core texts are fully and fluently translated. Domestication is the most intensively used approach in translating the military concepts and principles. At the same time, the concise and vivid style of the text is rather faithfully maintained with most of the metaphoric expression foreignized. Both TT1 and TT2 are acceptable and elegant, with TT2 being more accurate in strategic principles.

However, there still exist some deficiencies in the translated core texts that may undermine the reconstruction of the military classic. The alien flavor and subtle nuances of the Chinese strategic culture is not fully retained, nor is the profundity of Sun Tzu's military concepts and principles sufficiently carried over. In addition, the translated core texts lack interpretation and substantiation of Sun Tzu's abstract military ideas.

These deficiencies in the reconstruction of ancient Chinese strategic culture are compensated to a large extent by the paratexts of TT1 and TT2. Some paratexts are translated selectively from among the SL ones, while others are added newly by the translators or by prestigious third-party authors. These paratexts work in partnership with the core texts and aid in the reconstruction of strategic culture by means of supplementation, recontextualization, comparison and evaluation.

In both TT1 and TT2 paratexts, supplementation for Chinese strategic culture covers mainly three aspects: the interpretation and substantiation of Sun Tzu's strategic ideas in the core text, the introduction to the author Sun Tzu, as well as the influence of the text. Battles led by well-known generals in China are furnished to substantiate Sun Tzu's abstract theorization. Comparatively speaking, TT1 stresses more on the SL paratexts; while TT2 puts in more paratexts to highlight the impact of Sun Tzu around the globe. It has to be noted that TT1 stresses on the peace-orientation of Sun Tzu, while TT2 focuses more on the modern application of strategic principles.

Recontextualization is another important measure taken by TT2 to re-establish the status of ancient Chinese strategic culture. In some newly added paratexts the ancient text *AOW* is put into the contemporary global setting and some of Sun Tzu's military concepts are re-defined with Western strategic terms. In this way, Sun Tzu, an ancient Chinese strategist, has been turned into a modernized and globalized military thinker with great appeal to modern Western readers.

Comparison is frequently used in both new paratexts of TT1 and TT2 to set up a framework in which ancient Chinese culture can be positioned. In TT1, comparison is done to identify similarities between Sun Tzu and 22 Western strategists and between *AOW* and 13 Western military books. Many Western military episodes are used to foreground the truth in Sun Tzu's thinking. As a result, the image of a respectable and prestigious military sage is re-established for target readers. In TT2 paratexts, however, the contrast is done mainly between *AOW* and *On War* with an emphasis on differences. In Hart's view, *AOW* is realistic, moderate, clear, profound and fresh; while *On War* is logical, extreme, abstract, obscure and outdated. Such a comparison is helpful in setting up the canonical position of Sun Tzu in the Western strategic arena.

Evaluation is another important measure which the paratexts take to build up the status of Sun Tzu. In both TT1 and TT2, a large number of commendatory remarks are used to foster an image of military canon. In TT1, the translator Giles himself implants his praises into the "Introduction" and more than 20 notes. In TT2, applauding remarks are not only pronounced by the translator but also by Liddell Hart, the highly esteemed English strategist whose evaluation would be viewed as expert, objective and prestigious by the Western audience.

To sum up, by means of domestication and foreignization in core texts, as well as supplementation, recontextualization, comparison and evaluation in the paratexts, both TT1 and TT2 provide readers a panoramic picture of ancient Chinese strategic culture. The image of a venerated strategist and a true, profound, time-honored and prestigious military classic is re-established for the target readers, so they can become impressed by and receptive to it.

References

Ball, Dyer. 1910. "Sun Tzŭ on the Art of War." *Journal of the Royal Asiatic Society of Great Britain & Ireland* 42 (3): 961.

Bartholomees, J. Boone., ed. 2010. *The US Army War College Guide to National Security Issues Volume I: Theory of War and Strategy.* 4th ed. Carlisle: The US Army War College.

Collins, John M. 1973. *Grand Strategy Principles and Practices*. Annapolis, MD: Naval Institute Press.

Fu, Chao 傅朝. 2001. "A Research on Military Terms in *The Art of War* 《孙子兵法》军语研究." *Journal of Jinzhou Teachers College (Social Science Edition)* 锦州师范学院学报(哲学社会科学版) 23 (2): 51–52.

Giles, Lionel. trans. 1910. *On the Art of War: The Oldest Military Treatise in the World (Translated from the Chinese with Introduction and Critical Notes by Lionel Giles)*. London: Luzac.

Griffith, Samuel B. trans. 1963. *The Art of War (Preface by Liddell Hart)*. New York: Oxford University Press.

Guo, Huaruo 郭化若. 1984. *Sun Tzu Interpreted and Annotated* 孙子译注. Shanghai 上海: Shanghai Chinese Classic Publishing House 上海古籍出版社.

Handel, Michael I. 2001. *Masters of War: Classical Strategic Thought*. 3rd ed. London and Portland: Frank Cass.

Hart, Liddell. 1954. *Strategy: The Indirect Approach*. New York: Frederick A. Praeger.

Hart, Liddell. 1963. "Foreword." In *Sun Tzu: The Art of War*, trans. Samuel B. Griffith, vi–vii. New York: Oxford University Press.

Li, Ling 李零. 1991. *The Art of War with Interpretations and Annotations* 孙子兵法译注. Chengdu 成都: Bashu Press 巴蜀书社.

Minford, John. 2008. "Forward." In: *Sun Tzu The Art of War (Bilingual Edition with Complete Chinese and English Text)*, trans. Lionel Giles." Tokyo, Rutland, VT and Singapore: Tuttle Publishing.

O'Dowd, Edward, and Arthur Waldron. 1991. "Sun Tzu for Strategists." *Comparative Strategy* 10 (1): 25–36.

Sawyer, Ralph D. trans. 1993. *The Seven Military Classics of Ancient China (with a commentary)*. Boulder, CO: Westview Press.

Tao, Hanzhang. 2007. *Sun Tzu's Art of War: The Modern Chinese Interpretation,* trans. Yuan Shibing. New York: Sterling Publishing Company.

von Moltke, Helmuth G. 1893. *The Franco-German War of 1870–71*(Forbes Archibald. translated). London: James R. Osgood, McIlvaine & Co.

Wawro, Geoffrey. 2005. *The Franco-Prussian War: The German Conquest of France in 1870–1871*. New York: Cambridge University Press.

Zhou, Hengxiang 周亨祥. 1992. *Sun Tzu's The Art of War with Full Annotation* 孙子全译. Guiyang 贵阳: 贵州人民出版社.

5 Reception of translated Sun Tzu in Western discourse

A re-canonization process

In the previous Chapter, it was found that through a combination of approaches and methods in both the core texts and paratexts of their translations, Giles and Griffith provided their readers a panoramic picture of ancient Chinese strategic culture and reconstructed an image of a venerated strategist and a true, profound, time-honored and prestigious military classic for the target readers. This chapter will continue to give an account of how the translated Sun Tzu has been received in the Western military and non-military discourse, based on two self-built corpora of more than 490 texts dating from 1910 to 2020. We generally follow the analytical structure of culture at individual, institutional and societal levels. Section 5.1 presents an introduction to the general trends of reception of *AOW* translation in Western military discourse. Section 5.2, 5.3 and 5.4 present respectively the findings about the reception of *AOW* by individual strategists, by Western (especially US) military universities and forces, and in the non-military sphere.

5.1 An overview of the reception of translated Sun Tzu in Western military discourse

The reception of the translated Sun Tzu to the West started 117 years ago when the first English translation came into publication in 1905. This section tries to provide an overview of the reception of Sun Tzu in Western military discourse by addressing issues such as the diachronic change of the reception of *AOW* translations, the attitude with which the translations are received, the frequency with which they are quoted, the most popular translations and the most welcomed strategic principles of Sun Tzu. Most of the statistics are gathered with help of the corpus search software Antconc 3.4.3.

Firstly, the translated *AOW* has been gaining increasing academic interest and public popularity in Western strategic culture ever since the end of the Second World War. This can be inferred from the diachronic changes of quotation of the translated Sun Tzu based on the survey of the 302 military texts in our corpus (see Figure 5.1). The result of the survey shows that from 1910 to 1959, the translated Sun Tzu was rarely quoted with less than 1 text each year. From 1960 to 1989, on average 1.5 Western military texts quoted Sun

DOI: 10.4324/9781003025726-5

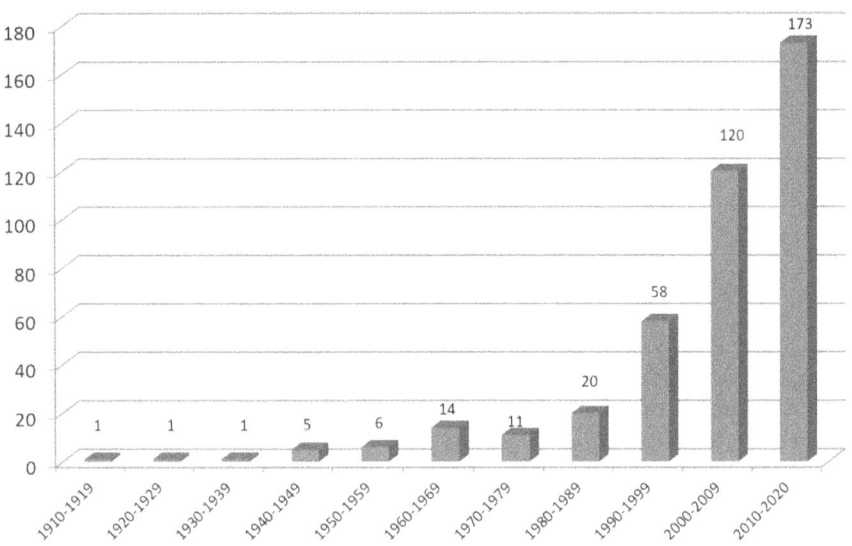

Figure 5.1 Diachronic change of the number of texts quoting translated Sun Tzu (1910–2020).

Tzu in each year. In the next decade from 1990 to 1999, the average number of texts quoting Sun Tzu increased to more than 5 each year. The annual quotes from 2000 to 2009 increased to 12. From 2010 to 2020, texts quoting Sun Tzu soared up to 15 annually.

Secondly, the attitude with which the translated Sun Tzu is received is positive. Our investigation shows that the overwhelming majority of the texts in Corpus 2, 99% of them, acclaimed that Sun Tzu's military wisdom is remarkable, profound, timeless and relevant to modern warfare (e.g. Lord 2000; McNeilly and McNeilly 2001; TUSI 1963). Among these 410 texts, only one criticizes Sun Tzu's idea of "not attacking cities" as "bad advice" (Leonhard 2003), two other texts propose that Sun Tzu's ideas are misappropriated or misunderstood (Hwang and Ling 2009; Ullman 2003 April 8).

Thirdly, there is a high frequency of direct quotations in Western military texts which shows that the translations of *AOW* have attracted intense attention in strategic research and military application. Among the 410 texts, 64 are headed with titles including "Sun Tzu" or "*The Art of War*". About 352 (86%) of them quote directly from the core texts of the translations (see Table 5.1), with an average number of five direct quotes in each text. For monographs and research reports, the average number of direct quotes per text is more than ten, which suggests that *AOW* translations are more intensively quoted in monographs and research reports.

Fourthly, paratexts from *AOW* translations also attracted close attention of strategic researchers. Our statistics show that 57 of the 410 texts have quoted

Table 5.1 Direct quotations from core texts in *AOW* translations

Text Type	Texts	Texts with direct quotes	Percentage	Number of direct quotes	Average number
Journal papers	190	166	87.36%	620	3.26
Monographs	58	51	87.93%	755	13.01
Military doctrines	47	42	89.36%	73	1.74
Book chapters	39	30	76.92%	116	2.97
Research reports	27	25	90.26%	306	11.33
Theses	25	18	72%	149	5.96
News articles & Online articles	24	20	83.33%	58	2.41
Total	410	352	85.85%	2077	5.07

Table 5.2 Quotations from paratexts in *AOW* translations

Text Type	Texts	Texts with quotes	Percentage	Number of quotes	Average number
Journal papers	190	18	9.47%	50	0.26
Monographs	58	16	27.58%	247	4.25
Military doctrines	47	0	0.0%	0	0.0
Book chapters	39	5	12.82%	10	0.26
Research reports	27	10	37.04%	23	0.85
Theses	25	6	33.3%	15	0.60
News articles & Online articles	24	2	8.33%	3	0.13
Total	410	57	13.90%	348	0.85

348 times from the paratexts of *AOW* translations, with an average number of less than one quote in each text (see Table 5.2). In monographs, which often involve detailed and lengthy discussions on strategic affairs, paratexts from *AOW* translations are most frequently quoted, with an average number of four quotes each.

The fifth issue concerns which translations are most welcomed in Western military discourse. According to our survey on *AOW* translations quoted in Corpus 2, the most popular translations is Griffith's, followed by Giles' (see Figure 5.2). Griffith's translation is quoted among 204 texts, taking up 45% of the total, Giles' translation is quoted among 102 texts, taking up 23% of the total. The third most frequently quoted translation is Sawyer's, followed closely by Ames'. It has to be noted that in some cases, a single text quotes two or more translations.

The sixth issue involves the most popular axioms and military principles of Sun Tzu quoted in Western military discourse. For this purpose, the direct quotes from Giles' and Griffith's translation in the 410 texts are

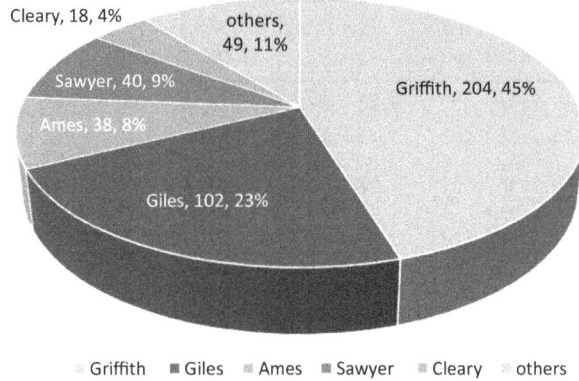

Figure 5.2 AOW translations quoted in Corpus 2.

surveyed and the result is shown in Table 5.3. According to the survey, the most popular Sun Tzu maxims are "For to win one hundred victories in one hundred battles is not the acme of skill. To subdue the enemy without fighting is the acme of skill" and "All warfare is based on deception". The most frequently quoted principles from the translated Sun Tzu are: (1) deception, (2) victory without fighting, (3) foreknowledge and information, (4) dialectical view and (5) holistic approach. In Corpus 2 there is almost one text out of every four quoting Sun Tzu's principle of "deception" or "victory without fighting".

In the following sections, we are going to investigate how the translated strategic wisdom of Sun Tzu found its way into Western strategic culture. Our purpose here is to find out who are quoting *AOW* translations in Western military and non-military discourse, and to what effect. We aim to map out the process of Sun Tzu's reception and incorporation of his concepts into Western culture. The analysis is conducted at three stages: the individual, institutional and societal levels.

5.2 At the individual level

Like many successful exchanges between cultures, the import of Sun Tzu's strategic thoughts into Western military culture was initiated by translations which were afterwards quoted, applied and developed in discourses by some far-sighted individuals including strategic theorists, military officers and politicians. Some of them have become Western "successors" to Sun Tzu, such as Liddell Hart and John Boyd (Yuen 2014, 125). Their personal contributions are evident among journal papers, monographs, research reports and even lectures. They are called "successors" not only because their innovative strategic ideas contain certain elements of Sun Tzu's thought, but also because "their

Table 5.3 Most quoted maxims and principles from translated Sun Tzu

Sun Tzu's principles	Version	Direct quote	Number	Sum
(1) Deception	Giles	All warfare is based on deception.	28	112
	Griffith	All warfare is based on deception.	55	
	Giles	Hence, when able to attack, we must seem unable; when using our forces, we must seem inactive; when we are near, we must make the enemy believe we are far away; when far away, we must make him believe we are near.	13	
	Griffith	Therefore, when capable, feign incapacity; when active, inactivity. When near, make it appear that you are far away; when far away, that you are near.	16	
(2) Victory without fighting	Giles	Hence to fight and conquer in all your battles is not supreme excellence; supreme excellence consists in breaking the enemy's resistance without fighting.	17	106
	Griffith	For to win one hundred victories in one hundred battles is not the acme of skill. To subdue the enemy without fighting is the acme of skill.	58	
	Giles	Thus, the highest form of generalship is to baulk the enemy's plans; the next best is to prevent the junction of the enemy's forces; the next in order is to attack the enemy's army in the field; and the worst policy of all is to besiege walled cities.	7	
	Griffith	Thus, what is of supreme importance in war is to attack the enemy's strategy, Next best is to disrupt his alliances. The next best is to attack his army. The worst policy is to attack cities.	24	
(3) Foreknowledge and information	Giles	Hence the saying: If you know the enemy and know yourself, you need not fear the result of a hundred battles. If you know yourself but not the enemy, for every victory gained you will also suffer a defeat. If you know neither the enemy nor yourself, you will succumb in every battle.	20	84`
	Griffith	Therefore I say: Know the enemy and know yourself; in a hundred battles you will never be in peril. When you are ignorant of the enemy but know yourself, your chances of winning or losing are equal. If ignorant both of your enemy and of yourself, you are certain in every battle to be in peril."	39	

(continued)

Table 5.3 Cont.

Sun Tzu's principles	Version	Direct quote	Number	Sum
	Giles	Hence the saying: If you know the enemy and know yourself, your victory will not stand in doubt; if you know Heaven and know Earth, you may make your victory complete.	2	
	Griffith	And therefore I say: "Know the enemy, know yourself; your victory will never be endangered. Know the ground, know the weather; your victory will then be total."	10	
	Giles	Now this foreknowledge cannot be elicited from spirits; it cannot be obtained inductively from experience, nor by any deductive calculation.	3	
	Griffith	What is called "foreknowledge" cannot be elicited from spirits, nor from gods, nor by analogy with past events, nor from calculations.	10	
(4) Holistic approach	Giles	The art of war is of vital importance to the State. Hence it is a subject of inquiry which can on no account be neglected.	14	66
	Griffith	War is a matter of vital importance to the State; the province of life or death; the road to survival or ruin. It is mandatory that it be thoroughly studied.	29	
	Giles	There is no instance of a country having benefited from prolonged warfare.	5	
	Griffith	For there has never been a protracted war from which a country has benefited.	18	
(5) Dialectical view	Giles	To ensure that your whole host may withstand the brunt of the enemy's attack and remain unshaken — this is effected by manoeuvres direct and indirect.	1	67
	Griffith	That the army is certain to sustain the enemy's attack without suffering defeat is due to operations of the extraordinary and the normal forces,	4	
	Giles	In all fighting, the direct method may be used for joining battle, but indirect methods will be needed in order to secure victory.	5	
	Griffith	Generally, in battle, use the normal force to engage; use the extraordinary to win.	6	
	Giles	Indirect tactics, efficiently applied, are inexhaustible as Heaven and Earth, unending as the flow of rivers and streams.	5	

Table 5.3 Cont.

Sun Tzu's principles	Version	Direct quote	Number	Sum
	Griffith	Now the resources of those skilled in the use of extraordinary forces are as infinite as the heavens and earth; as inexhaustible as the flow of the great rivers.	4	
	Giles	In battle, there are not more than two methods of attack — the direct and the indirect; yet these two in combination give rise to an endless series of manoeuvres. The direct and the indirect lead on to each other in turn.	6	
	Griffith	In battle there are only the normal and extraordinary forces, but their combinations are limitless; none can comprehend them all.	9	
	Giles	It is like moving in a circle — you never come to an end. Who can exhaust the possibilities of their combination?	3	
	Griffith	For these two forces are mutually reproductive; their interaction as endless as that of interlocked rings. Who can determine where one ends and the other begins?	8	
	Giles	Therefore, just as water retains no constant shape, so in warfare there are no constant conditions.	5	
	Griffith	And as water has no constant form, there are in war no constant conditions.	11	

attempts to redefine and re-theorize Western strategy have made Western strategic thought more attuned to Sun Tzu's thought as well as Chinese strategic thought as a whole" (Yuen 2014, 125). In the following paragraphs, individual discourses involving some of these key "successors" to Sun Tzu will be discussed.

5.2.1 *Liddell Hart: from* cheng *and* ch'i *to indirect approach*

Liddell Hart, one of the most prominent strategic thinkers of the twentieth century, was the first one to accommodate Western strategic thought to Sun Tzu's ideas. His two major contributions were his development of "indirect approach" based on Sun Tzu's dual concepts of *ch'i* and *cheng*, as well as the establishment of the concept and framework of "grand strategy" in Western strategic thought. Both of them have absorbed inspirations from Giles' translation of *AOW*.

In the spring of 1927, Hart's interest in *AOW* "was aroused by a letter" he received from a friend, which mentioned two strategic ideas from Sun Tzu (Hart 1963, vi). One strategic idea was expressed by Giles' faithful translation of Sun Tzu in the water simile "military tactics are like unto water". The other was the principle of winning without fighting. Prior to this letter, Hart had formulated his own "Expanding Torrent Theory" of attack, which features an automatic and continuous progressive infiltration by combat units. Later he recalled that: "when I read Sun Tzu's two-thousand-year-old book on *The Art of War*, I found that he had used a close simile" (Hart 1954, 44). Encouraged by the resonance between Sun Tzu's "water simile" and his own "Torrent Theory", Hart continued to digest what is embodied in Giles' translation and use Sun Tzu's ideas to develop his own strategic theory. This case reveals the unique role of faithful translation of Sun Tzu's metaphoric expressions in arousing readers' interest.

Hart's most reputed theory "Strategy of the Indirect Approach" emerged in 1928 and was further expounded in his book *The Decisive Wars of History* (Hart 1929). This book systematically analyses the history of wars from ancient Greece through the nineteenth century and puts forwards a theoretical discussion of indirect strategy at the conclusion. The later version of the book was extended to cover the Second World War and published with a title more accurately expressing Hart's idea, *Strategy: The Indirect Approach* (Hart 1954). Hart's indirect approach "represents the first systematic effort to synthesize Chinese and Western strategic thought" (Yuen 2014, 128–129), which can be judged from the following aspects.

On the flyleaf of *Strategy: The Indirect Approach,* there are 20 quotes concerning the art of strategy (See Appendix 5). Among these quotes, 14 (amounting to 70% of the total) are from Giles' translation of Sun Tzu, while the rest are from six famous historical figures: Belisarius, Shakespeare, Napoleon, Clausewitz, Moltke and Admiral De Robeck. These quotes are pointing to Sun Tzu's five principles of deception, victory without fighting, dialectical view (*cheng* and *ch'i*), etc. This implies that when conceiving his indirect strategy, Hart followed the traditions of Western strategy, but more frequently followed the heritage of Chinese strategy.

It can be easily spotted that inside the book, Hart's juxtaposition of indirect and direct approaches replicates Sun Tzu's dual-concepts of *ch'i* and *cheng* (Yuen 2014, 127). According to Hart, the art of the indirect approach can be crystallized into two simple maxims: "in face of the overwhelming evidence of history, no general is justified in launching his troops to a direct attack upon an enemy firmly in position" and "instead of seeking to upset the enemy's equilibrium by one's attack, it must be upset before a real attack is, or can be successfully launched" (Hart 1954, 164). Further, a strategic indirect approach "may start by being indirect in relation to the enemy's front, but by the very directness of its progress towards his rear may allow him to change his dispositions, so that it soon becomes a direct approach to his new front" (Hart 1954, 34).

Hart's statement about the aim of indirect strategy sounds more Eastern: "[f]or even if a decisive battle be the goal, the aim of strategy must be to bring about this battle under the most advantageous circumstances. And the more advantageous the circumstances, the less, proportionately, will be the fighting" (Hart 1954, 339). This statement is opposed to his contemporary mainstream of military paradigm which aimed for the destruction of the enemy's forces by means of a decisive battle.

Like Sun Tzu, Hart stressed on movement, flexibility and surprise:

> Strategy has not to overcome resistance, except from nature. *Its purpose is to diminish the possibility of resistance*, and it seeks to fulfill this purpose by exploiting the elements of *movement* and *surprise*... Movement lies in the physical sphere, and depends on a calculation of the conditions of time, topography, and transport capacity... Surprise lies in the psychological sphere and depends on a calculation, far more difficult than in the physical sphere, of the manifold conditions, varying in each case, which are likely to affect the will of the opponent.
>
> (Hart 1954, 337)

Prior to Hart, grand strategy had not been formulated and systematically studied in the West. Hart's concept of grand strategy is "distinctively Chinese in at least three ways" (Yuen 2014, 133). In the first place, while imitating Sun Tzu, Hart extended the concept of the indirect approach from the tactical level to the level of grand strategy. For Hart, grand strategy should both calculate and develop the economic resources, manpower of nations and the moral resources.

> Fighting power is but one of the instruments of grand strategy—which should take account of and apply the power of financial pressure, of diplomatic pressure, of commercial pressure, and, not least of ethical pressure, to weaken the opponent's will.
>
> (Hart 1954, 336)

In a similar vein to Sun Tzu, Hart's indirect approach embraces both physical and psychological dimensions, as he said: "in studying the physical aspect we must never lose sight of the psychological, and only when both are combined is the strategy truly an indirect approach, calculated to dislocate the opponent's balance" (Hart 1954, 34).

Secondly, Hart's concept of grand strategy draws heavily upon Sun Tzu's principle of winning without fighting for its theoretical basis. One sentence about Hart's ideal strategy reads: "the perfection of strategy would be, therefore, to produce a decision without any serious fighting" (Hart 1954, 339). Obviously, this remark is nothing but a rewording of Giles' translation of Sun Tzu's principle of victory without fighting. This maxim is extremely important to Hart as it articulates the possibility of victory without fighting at the tactical level and suggests the use of non-military means so that Hart was inspired to upgrade "[W]estern strategy to the height of grand strategy" (Yuen 2014, 135).

Lastly, the peace-oriented thinking in Hart's grand strategy concept is almost perfectly in line with that in *AOW*. Hart (1954, 362) distinguished the fundamental difference between strategy and grand strategy: "[w]hereas strategy is only concerned with the problem of winning military victory, grand strategy must take the longer view—for its problem is the winning of the peace". Furthermore, Hart explained that:

> Victory in the true sense implies that the state of peace, and of one's people, is better after the war than before. Victory in this sense is only possible if a quick result can be gained or if a long effort can be economically proportioned to the national resources…Peace through stalemate, based on a coincident recognition by each side of the opponent's strength, is at least preferable to peace through common exhaustion—and has often provided a better foundation for lasting peace.
>
> (Hart 1954, 370)

Hart not only quoted Giles' translation to cultivate his own theory but also helped to spread the influence of Griffith's new translation of Sun Tzu. In 1963, he authored the *Foreword* to Griffith's translation, and acclaimed that *AOW* embodies "the concentrated essence of wisdom on the conduct of war". He also acknowledged his inheritance from Sun Tzu's theory:

> I found many other points that coincided with my own lines of thought, especially his constant emphasis on doing the unexpected and pursuing the indirect approach. It helped me to realize the agelessness of the more fundamental military ideas, even of a tactical nature.
>
> (Hart 1963, vi–vii)

Although Liddell Hart's adoption of Sun Tzu's strategic thinking was selective and partial, it "clearly demonstrated the validity of Chinese strategic thought and its broader applicability to modern Western settings" (Yuen 2014, 2). Thanks to the works of Liddell Hart, *AOW* began to have a greater impact on Western strategic thought and embarked on the way for further incorporation into its Western counterpart (Yuen 2014, 125–126).

The above analysis shows that Hart's term "indirect approach" is basically borrowed from the domesticating translation of Sun Tzu's concept of *ch'i*. Later, Western strategists began to accept the transliteration of the key military concepts *chéng* and *ch'i* initiated by Giles in his paratext. O'Dowd and Waldron (1991, 30) stated that *ch'i* and *chéng* are among the fundamental elements of Sun Tzu's military thoughts. According to Lord (2000, 305), Sun Tzu's key strategic concepts, such as *chéng* and *ch'i*, are of great value, precisely because they emerged in an alien cultural matrix and yet retain a compelling logic for contemporary readers, and they are, in some respects, superior at the level of pure strategic theory to corresponding Western notions. Lord (2000) also discussed the terms *hsing* (形, meaning form or position) and *shih* (势, meaning advantage or leverage), which are taken from the transliteration of Sun Tzu's concepts.

5.2.2 Michael Handel: comparing Sun Tzu with Western strategists

In order to identify the similarities and differences between Eastern and Western strategic thinking and culture, many military theorists have compared Sun Tzu with Clausewitz, Corbett or Mahan (Fitzsimmons 2007; Handel 1991; 1992; 2000; 2001; Paquette 1991). On the one hand, their comparisons are often based on extensive quotes from translations of Sun Tzu. On the other hand, their comparisons can be viewed as the continuation of the comparison initiated by translators such as Giles and Griffith in their paratexts as mentioned in Chapter 4. Among these comparing experts, Michael Handel is a typical example.

Michael Handel, with a PhD degree at Harvard University, was an expert on strategic theory, nature and operations of war, and the future of warfare. He was Professor of National Security Affairs in the US Army War College from 1983 to 1990 and a Professor of Naval Strategy in the US Naval War College from 1990 to his death in 2001. As a member of the Olin Institute for Strategic Studies at the Center for International Affairs in Harvard University and founder and editor of the journal *Intelligence and National Security*, he authored many books on theory and practice of war.

In his monograph published in 2001 entitled *Masters of War: Classical Strategic Thought* (3rd revised and expanded edition), Handel compared *AOW* and Carl von Clausewitz's *On War*, two enduring classic military books where the tradition of Eastern and Western military strategies can be traced. Handel (2001, 2–5) believed that *AOW* and *On War* are largely in agreement on the fundamental issues, such as a rational analysis of war and sufficient freedom afforded to the field commander. However, it is undoubted that they are different in many aspects in the views on waging war, the role of force, the ideal victory and preferred way of winning (Handel 2001, 6). *AOW* holds a broad perspective that includes a great variety of non-military means (e.g., diplomatic, economic and psychological), while *On War* has a narrow emphasis on the use of military means. In *AOW*, military force should be used frugally and as the last resort; while in *On War*, the use of force is often necessary and the most effective (or preferred) method to achieve the political goals of the state. The maximum available force should be used from the beginning to achieve decisive results in the shortest possible time. In Sun Tzu's view, the best way is to win without fighting, either through diplomatic means before war breaks out, or by convincing the enemy's forces to yield. However, in Clausewitz's principle of destruction, the shortest way to achieve one's political objectives is the destruction of the enemy's forces in a major battle. In *AOW,* extensive use of deception and psychological warfare are acclaimed with the centre of gravity laid on the enemy's will and alliance system; whereas in *On War*, the maximum concentration of force is at the decisive point of engagement, with the centre of gravity on the enemy's army.

Handel's comparison of *AOW* and *On War* is a systematic and in-depth continuation of Giles' comparative notes and Liddell Hart's comparative remarks in the preface to Griffith's translation. For a strategy theorist who was not familiar with Chinese language, Handel's comparison relies heavily on the translations. He quoted 171 times from the translations of Sun Tzu's core texts and 55 times from the translated paratexts. Although most of the quotations come from Griffith's translation, Handel also consulted and compared from many other translations so as to find out the truest and most appropriate interpretation of Sun Tzu's maxims. Altogether 18 versions of the translated Sun Tzu are listed in the bibliography of his book.

One specific example involves Handel's research on Sun Tzu's concept of *shih*, or comparative strategic advantages (Handel 2001, 84–88). According to Sun Tzu, the art of strategy should be based on exploiting a comparative advantage that enables one to fight on his own best terms. To expound Sun Tzu's concept, Handel listed fours translations respectively by Lionel Giles, Samuel Griffith, Roger Ames and Tai Mien-leng. In addition, he cited a commentary by Tu Mu from Giles' notes for further clarification: "[o]ne mark of a great soldier [or strategist] is that he fights on his own terms or fights not at all" (Giles 1910, 42). In the following paragraphs, Handel continued to quote directly five times from Griffith translation to expound on how to gain comparative advantage. These quotations were grouped together from various parts of *AOW* so as to form a complete cohesive link of "the process of thinking, planning, and searching for a comparative advantage" (Handel 2001, 86). Based on an understanding gained from the effort of comparing translations, Handel was able to offer an in-depth comparison between Sun Tzu and Clausewitz in the aspect of comparative advantage. He discovered that Clausewitz did not discuss the need to search for a comparative advantage as the basis for a sound strategy. Finally, Handel concluded that "Sun Tzu makes an important and original contribution to the study of war when he explicitly develops the concept of comparative advantage" (Handel 2001, 86).

In another case, Handel reflected on Sun Tzu's maxim of victory without fighting: "故上兵伐谋，其次伐交，其次伐兵，其下攻城。" In Griffith's translation, it goes "Thus, what is of supreme importance in war is to attack the enemy's strategy, Next best is to disrupt his alliances. The next best is to attack his army. The worst policy is to attack cities" (Griffith 1963, 182–183). Handel believed that a better understanding of this proverb could be gained by "(a) comparing Griffith's translation with a number of others; and (b) viewing the statement and the phrase following it in the context of Sun Tzu's discussion" (Handel 2001, 296). For this purpose, he composed a note running more than three pages in about 1600 words (Handel 2001, 296–300). In the note, he listed ten versions of the translated proverb by other nine translators: Giles, Cleary, Huang, Chen, Ames, Sawyer, Tang, Cheng and Tai. In addition, he also cited a translated paratext from Giles' translation.

Handel (2001, 296) discovered that when translating the phrase "伐交", Giles regarded it as a tactic at the operational level as he translated the phrase into "prevent the junction of the enemy's forces", while other translators agreed that it is at the national strategic level to use diplomatic measures to undermine the enemy's alliances. In translating "谋", translators used different words such as "plans", "strategy" and "policy", consequently putting the concept of "谋" at different levels. The word "plan", suggests a specific preparatory activity using precise knowledge regarding the enemy's intentions; "military strategy" may suggest the use of violent force; while "policy" includes violent and non-violent means. Handel realized that most of the translations indicate that "谋" should be implemented before the outbreak of war, while Griffith translation "stratagem" suggests that this is the best course of action in war. These different translations bring about two different interpretations of Sun Tzu. One is the simplest interpretation that Sun Tzu is referring to a purely military activity. The other is a more common interpretation that it is a policy of attacking the enemy's strategy, plans or intentions by non-military means.

Handel (2001, 297) preferred the latter interpretation: "Sun Tzu favors the use of diplomacy and other non-violent approaches—not just military action". He believed that Sun Tzu presented a four-stage progression, "a continuum of action from peace to war, from non-violent means to the use of force and from the most cost-effective to the least cost-effective military methods of attaining one's objectives" (Handel 2001, 299). Particularly, attacking the enemy's strategy is the first stage, which aims to win at the lowest possible cost. It refers to a pre-war phase in which non-military methods of achieving the state's goals are given precedence.

With an understanding gained from the comparative analysis of different translations, Handel concluded that Sun Tzu conducted a higher-level analysis blending diplomatic (and other) means with military methods. While Clausewitz pondered exclusively on war, Sun Tzu deliberated on the art of war, diplomacy and statecraft. Handel (2001, 299) pointed out that "[i]n the West, such pre-war policies and the use of non-military means would be considered grand strategy or statesmanship". Some experts argued that "the pre-modern logic of Sun Tzu makes a better guide to late-modern military thinking than the modern logic of Clausewitz" (Rasmussen 2001, 25).

5.2.3 Richard Nixon: citing Sun Tzu for US grand strategy

Like Giles' translation, which had a profound impact on the great strategist Liddell Hart, Griffith's translation also became a great source of inspiration for many other strategists such as Richard Nixon and John Boyd.

Richard Nixon (1913–1995) was the 37th US president, who served from 1969 to 1974 at a time of upheaval and change, with a clear vision of grand strategy mostly likely influenced by the wisdom of Sun Tzu. He withdrew American troops in 1973 from their protracted involvement in the Vietnam War and improved international relations with the Soviet Union and China. His visit to China in 1972 started the Sino-US diplomatic relations. In the same year, after the intense negotiations following his summit meetings with Russian leaders, two landmark arms control treaties were signed to address the potential catastrophic risk of nuclear war: the Strategic Arms Limitation Talks and the Anti-Ballistic Missile Treaty.

Nixon authored several books after his resignation from the presidency in 1974. One of them is *The Real War*, a landmark 1980 bestseller. This book presents a shrewd analysis of US strategic shortcomings and a prescription for US to use the political, economic, military strengths and national will to turn the tide: "we must drastically increase our military power, reinvigorate our willpower, strengthen the power of our Presidents and develop a strategy aimed not just at avoiding defeat but at attaining victory" (Nixon 1980, 15). His strategic views became a blueprint for Ronald Reagan's military and strategic initiatives, which finally paved the way for the end of the Cold War.

Inside this book, Nixon directly cited Griffith's translation of Sun Tzu three times to discuss strategic issues in addition to many indirect quotations to it. When commenting on the American action in Vietnam, he quoted Griffith's translation to criticize the prolonged US involvement in war and to illustrate its negative effects: "[t]here has never been a protracted war from which a country has benefited... What is essential in war is victory, not prolonged operations" (Nixon 1980, 105).

While discussing the US-Soviet strategic superiority, Nixon used the phrases "victory [win] without war" five times, and "defeat without war" four times, which are obvious adaptations from translated Sun Tzu principles. For instance, he warned that the US "might be defeated without war" (Nixon 1980, 2), declared "a strategic advantage for the United States and the West reduces the danger of war or defeat without war" (Nixon 1980, 153), and remarked that "the crucial element in developing a strategy to win victory without war is willpower. Military power and economic power are necessary, but they are useless without willpower" (Nixon 1980, 247).

When analyzing the nuclear strategy of the two superpowers, Nixon listed the three faults with the concept of mutual assured destruction (MAD): the Soviet's refusal to go along with it, no rational political or military objectives, and a morally wrong threat. Therefore, the MAD thought should be abandoned. Nixon quoted the translation of Sun Tzu for support: "[w]hat is of supreme importance in war is to attack the enemy's strategy... Attack cities only when there is no alternative" (Nixon 1980, 162).

Nixon suggested that in order for the US to acquire victory, the US should "check Soviet strengths and exploit Soviet weaknesses". According to him, Russia was militarily superior, but economically and morally inferior at that time. To further explain his strategy, Nixon once again referred to Sun Tzu:

More than 2,000 years ago the ancient Chinese strategist Sun Tzu set forth this principle: Engage with the *ch'eng*—the ordinary, direct force— but win with the *ch'i*—the extraordinary, indirect force. In his wisdom he saw that the two are mutually reinforcing and that the way to victory is by the simultaneous use of both.

In our own time we have no choice but to engage with the *ch'eng*—to counterpose our military strength to that of the Soviet Union, to hold our alliances together and increase the combined strength of the West. This is the way to avoid defeat; this is the way to contain Soviet advance. It is an essential first step, just as the tide has to stop coming in before it goes out. The next step—to go on toward victory, to win with the *ch'i*—is at once more complex, more subtle, and more demanding. Yet here again the West has the greatest advantages, if only we can marshal and use them.

(Nixon 1980, 300)

Nixon's application of Sun Tzu's *ch'i and cheng* is at the grand strategy level, and such application brings about tremendous influence on the international security. His influential book *The Real War* has once again promoted the wisdom of Sun Tzu to Western strategists, military officers, diplomats and beyond. The case of *Real War* highlights the effectiveness of Griffith's coordinated approach of domesticating Sun Tzu's *ch'eng* and *ch'i* in the core text and foreignizing it in the paratexts.

5.2.4 *John Boyd: the American Sun Tzu and his OODA Loop Theory*

John Boyd (1927–1997) was a US Air Force fighter pilot and one of the premier military strategists of his time, whose theories had been tremendously influenced by Sun Tzu. Berkowitz (2003) named Boyd as "The American Sun Tzu", while Osinga (2007, 4) remarked that Sun Tzu was "Boyd's conceptual father". As a devoted disciple of Sun Tzu, Boyd spent many years researching the translated works of Chinese strategists, as his biography writer Robert Coram recorded:

The Art of War became Boyd's Rosetta stone, the work he returned to again and again. It is the only theoretical book on war that Boyd did not find fundamentally flawed. He eventually owned seven translations, each with long passages underlined and with copious marginalia. The translations of Samuel Griffith and, later, Thomas Cleary were his favorites.

(Coram 2002, 331)

Boyd was best known for two original achievements. The first one is the mathematically coherent theory of air combat, "energy maneuverability", to calculate the advantage in any flight state (the combination of airspeed, altitude and direction). This theory has become the worldwide standard for the design of fighter aircraft. The second contribution to modern strategy is the Observe-Orient-Decide-Act Loop (OODA Loop) developed on the basis of energy maneuverability to describe the process by which an entity (either an individual or an organization) reacts to an event. The four steps are: Observe (gather data about the situation), Orient (evaluate the data against existing knowledge and the objective), Decide (choose a course of action) and Act (execute the action). According to the loop, the essence to victory is to be able to create circumstances wherein one can make appropriate decisions faster than his opponent. This loop, with its direct link to the strategic purpose of Sun Tzu, is perhaps "the most brilliant insight of strategy in the last 100 years" (Richards 2003, 28).

Boyd did not publish much. However, his lectures with many slides of PowerPoint Presentations (PPT) to illustrate and examine the OODA theory have become a great influence upon his audience. These PPT presentations, collected under the title of *A Discourse on Winning and Losing*, including *Patterns of Conflict* (Boyd 1986), *Organic Design for Command and Control* (Boyd 1987a), and *The Strategic Game of? and?* (Boyd 1987b), In these presentations, Boyd reviewed many famous battles of reputed ancient and modern strategists, such as Clausewitz, Jomini, Napoleon, Saxe, Genghis Khan, Fuller and Ludendorff. Among them, "Sun Tzu was the only one that Boyd did not critique" (Richards 2003, 20), and Boyd "drew heavily on Sun Tzu's works in his examination of strategy" (Richards 2003, 17). More importantly, Boyd wrote "to convince people that the military doctrine and practice of his day were fundamentally flawed", and aimed for "an almost full adoption of Sun Tzu's thought into the Western strategic framework" (Yuen 2014, 137).

In the presentation *The Strategic Game of? and?* Boyd quoted the translated Sun Tzu: "Know your enemy and know yourself; in one hundred battles you will never be in peril" (from Giles' translation) and "[s]eize that which your adversary holds dear or values most highly; then he will conform to your desires" (from either Giles' or Griffith's translation). In another PPT, *Organic Design for Command and Control*, he indirectly quoted Sun Tzu: "Employ *cheng/ch'i* maneuvers to quickly and unexpectedly hurl strength against weaknesses."

The *Patterns of Conflict* is a 193-slide presentation for an 8-hour briefing, which Boyd developed over the course of about ten years (Richards 2003, 23). It embodies Boyd's core ideas on conflict and warfare and at the same time reflects his reading of military history "through the lens of Sun Tzu" (Yuen 2014, 137). This briefing starts with Sun Tzu and other ancient commanders, then takes the readers to commanders of the twentieth century and finally ends with Sun Tzu again. Boyd put the name of Sun Tzu in the titles of 146 to 156 of the slides, and borrowed frequently from Sun Tzu's ideas such as *cheng/ch'i*, the weak and strong in other slides (Osinga 2007, 196).

In the beginning, Sun Tzu's theme, strategy and desired outcome are listed as a benchmark to measure other ancient commanders, Alexander, Hannibal,

Belisarius, Genghis Khan and Tamerlane. Boyd directly and indirectly quoted the translated Sun Tzu in the following:

Strategy
   Probe enemy's organization and dispositions to unmask his strengths, weaknesses, patterns of movement and intentions.
   "Shape" enemy's perception of world to manipulate his plans and actions.
   Attack enemy's plans as best policy. Next best disrupt his alliances. Next best attack his army. Attack cities only when there is no alternative.
   Employ *cheng* and *ch'i* maneuvers to quickly and unexpectedly hurl strength against weaknesses.

Desired outcome
   subdue the enemy without fighting
   avoid protracted war

(Boyd 1986)

Boyd found that ancient commanders have at least one thing in common: action of *cheng* and *ch'i*. "*cheng/ch'i* maneuver schemes were employed by early commanders to expose adversary vulnerabilities and weaknesses (a la *cheng*) for exploitation and decisive stroke (via *ch'i*)" (Boyd 1986, 15). Boyd then summarized that: early commanders and Eastern ones seem consistent with ideas of Sun Tzu in attempting to shatter adversary prior to battle, while Western commanders were "more directly concerned with winning the battle" (Boyd 1986, 15). In the following slides, Boyd looked at an early tactical theme and some battle situations to understand the different ways that the *cheng/ch'i* game has been played.

Next, Boyd used Sun Tzu's theoretical framework to evaluate the two powerful types of warfare in the twentieth century at the two extremes of the scale— Blitzkrieg and guerilla warfare. Through the lens of Sun Tzu, the common conceptual foundation between these two contrasting ways of warfare is discovered:

   Blitz and guerrillas, by operating in a directed, yet more indistinct, more irregular, and quicker manner, operate inside their adversaries' observation- orientation-decision-action loops or get inside their mind-time-space as basis to penetrate the moral-mental-physical being of their adversaries in order to pull them apart, and bring about their collapse.

(Boyd 1986)

Boyd went on to put forward countermeasures against Blitzkrieg and guerrilla movement in light of Sun Tzu from three levels: strategy, tactics and grand maneuver. He also analyzed three kinds of conflicts: attrition warfare, maneuver conflict and moral conflict, and finally came up with his patterns for successful operations: shape or influence events, operate inside adversary's OODA loops and penetrate adversary's moral-mental-physical being.

Consequently, the basis for Boyd's moral-mental-physical conflict in three dimensions is established. Respectively, these dimensions deal with the destruction of the enemy's physical strength (fighting power), disorganization of his mental processes (thinking power), and disintegration of his moral will to resist (staying power) (Yuen 2014, 139).

With all the above-mentioned effort, Boyd succeeded in his "more extensive adoption of Sun Tzu's thesis into the Western strategic framework", and in "bridging the gap between Chinese and Western strategic thought" (Yuen 2014, 2–3).

Moreover, Boyd was a teacher who taught OODA Loop and at the same time spread earnestly Sun Tzu's strategic ideas. Boyd delivered his presentation *Patterns of Conflict* over 1,500 times during the 1980 and 1990s, in places like the Marine Basic School, the Army War College as well as the Command and Staff General College (Phua 2007, 47). His lectures made an impact on many listeners in the military sphere including Dick Cheney (the then Secretary of Defense), Senator Sam Nunn (former Chairman of the Senate Armed Services Committee), General Al Gray (former Commandant of USMC) and General Edward Myer (former Chief of Staff of the US Army). These lectures spread a better understanding of and more interest in Sun Tzu.

One of the consequences of Boyd's lectures was that they paved the way for the application of Sun Tzu to the First Gulf War. Boyd not only contributed theoretically, but also acted as "the chief architect behind the application of Sun Tzu in the Gulf War" (Phua 2007, 47), which will be discussed in Section 5.3.

The analysis reveals that translations of Sun Tzu have become the source from which Boyd learned Chinese strategic thoughts. Unlike Liddell Hart, he began to implant the transliteration of Sun Tzu's terms, such as *cheng* and *ch'i*, into his theorization of modern strategy and incorporation of Sun Tzu's ideas into Western strategic framework.

5.3 At the institutional level

With the concerted efforts of a large number of military officers, strategists and politicians who are researching, commenting and citing the translated Sun Tzu, Western military discourse began to embrace Sun Tzu at the institutional level mainly in three aspects: military doctrine, education and conflicts.

5.3.1 Sun Tzu in Western military doctrines

One of the most important signs of Sun Tzu's entry into Western military institutions is that *AOW* translations are cited in military doctrines. Military doctrines are fundamental rules by which military forces guide their actions in support of objectives, which are authoritative but require judgement in application (NATO Standardization Agency 2013, 2-D-9). A doctrine provides a common conceptual framework or a common lexicon for use by military planners and leaders. It helps standardize operations and provide the military with an authoritative body of statements on how military forces conduct operations to accomplish military tasks. Therefore, *AOW* translations' access into military doctrines undoubtedly marks a new stage of Sun Tzu's reception into Western strategic culture.

As our survey shows, at least 46 military doctrines from the US, UK, Canada and Australia have referred to Sun Tzu 68 times, and 41 (89.1% of the total) of them quoted Sun Tzu directly. Especially in the US, the translated Sun Tzu has been extensively quoted and has directly influenced the doctrines for all the four branches of the US military: Army, Navy, Air Force and Marine Corps.

For instance, the 1982 edition of the US Army's doctrine, *FM 100–5 Operations* (HDA 1982), considered as innovative in its theory of Air-Land cooperation as part of the Army's basic operational concepts, incorporates a fusion of the military theories of the past, especially those from Sun Tzu and Clausewitz. There are two direct quotations of Sun Tzu in it. Before going to the details of the basic tenets of successful Air-Land Battle (namely initiative, depth, agility and synchronization), a translation of Sun Tzu was quoted to provide support: "[r]apidity is the essence of war; take advantage of the enemy's unreadiness, make your way by unexpected routes, and attack unguarded spots" (HDA 1982, 2–1). Another quote appears when the doctrine is discussing urbanized terrain. It points out the importance of urban centers as strategic objectives and the difficulty in seizing cities and towns. It cites Sun Tzu: "the worst policy of all is to besiege walled cities". Then it continues to point out that many armies have learned the wisdom of Sun Tzu's words at such places as Stalingrad, Tobruk, Hue and Beirut (HDA 1982, 3–9).

The US Marine Corps are rather receptive to Sun Tzu. *FMFM 1 Warfighting* and *FMFM 1–1 Campaigning* are the two most fundamental marine doctrines that incorporate Sun Tzu's concepts. These doctrines were drafted by Captain John Schmitt under the direction of the Marine Corps Commandant, General Alfred M. Gray. Both Gray and Schmidt later insisted that *FMFM 1* was inspired primarily by Sun Tzu (Phua 2007, 47). Millett (1991, 634) holds that the "concise distillations of operational concepts [mostly Gray's] bore some resemblance to the works of Sun Tzu and Mao", and with a detailed comparison of the content of these two doctrines with *AOW,* Candela (1998, 255) discovers that "it may seem as though Marine Corps philosophy and doctrine is more similar to Sun Tzu's *The Art of War* than it is a direct product of Sun Tzu's *The Art of War*", and "it may appear that *The Art of War's* contributions are greater than most people recognize".

In *FMFM 1 Warfighting*, there are 12 quotations under the titles of four chapters (Headquarters US Marine Corps 1989). Griffith's translation of Sun Tzu is quoted three times, which is the largest number followed by two quotes from Clausewitz. The three quotes from *AOW* translation are about invincibility of the defence, the military advantage expressed by water simile, as well as speed and surprise attack. Furthermore, in a note for the second chapter "The Theory of War", it is recommended that

> [l]ike *On War*, *The Art of War* should be on every Marine officer's list of essential reading. Short and simple to read. *The Art of War* is every bit as valuable today as when it was written about 400 B.C.
>
> (Headquarters US Marine Corps 1989, 81)

FMFM 1–1 Campaigning (Headquarters US Marine Corps 1990) is another key manual for the US Marine Corps. It has four parts: chapter 1 "The Campaign", chapter 2 "Designing the Campaign", chapter 3 "Conducting the Campaign" and "Conclusion". Chapter 1 quotes Sun Tzu directly "a skilled commander seeks victory from the situation and does not demand it of his subordinates" when arguing that operations must serve tactics by creating the most advantageous conditions for tactical actions (Headquarters US Marine Corps 1990, 11). In chapter 2, it is remarked that the most effective way to defeat the enemy is to destroy an object of strategic importance, or to strike the enemy where and when they can be hurt most. Here, Griffith's translation of Sun Tzu is cited for support: "Seize something he cherishes and he will conform to your desires" (Headquarters US Marine Corps 1990, 36). In chapter 3, Sun Tzu's maxim of victory without fighting is put immediately under the title of the chapter (Headquarters US Marine Corps 1990, 53). The fourth quotation from Cleary's translation comes right after the title of the Conclusion: "[t]hose who know when to fight and when not to fight are victorious. …Those whose generals are able and are not constrained by their governments are victorious" (Headquarters US Marine Corps 1990, 85).

The translated Sun Tzu is also quoted in many US Air Force doctrines. For instance, in *Air Force Doctrine Document 2–5.3 Psychological Operations (Headquarters, US Air Force Doctrine Center 1999)*, two sentences from Giles' translation are quoted. One outlines Sun Tzu's principle of winning without fighting, put right after the head of the second chapter and used to highlight the importance of air force psychological operations (Headquarters, US Air Force Doctrine Center 1999, 7). The other summarizes the principle of deception, applied as a basis for the argument that "one effective means of gaining an advantage over the enemy is to build military deception into battle plans and individual missions" (Headquarters, US Air Force Doctrine Center 1999, 28).

Sun Tzu is also quoted in over 15 joint publications issued by the Joint Chiefs of Staff (JCS). JCS is an advisory body led by military service chiefs from the Army, Navy, Air Force, the Marine Corps and the chief of the National Guard Bureau, with a primary responsibility to ensure the personnel readiness, policy, planning and training of their respective military services for the combatant commanders to utilize. JCS also acts in a military advisory capacity for the US president and the Secretary of Defense. Joint Publications approved by JCS present fundamental principles that guide the employment of US military forces in coordinated and integrated action toward a common objective, including planning, training and conducting military operations. In the *Joint Pub 3–13 Joint Doctrine for Information Operations* (US Joint Chiefs of Staff 1998), Griffith's translation of Sun Tzu is quoted three times. Chapter 1 "Introduction" is headed by Sun Tzu's remarks on the principles of *cheng* and *ch'i*. In chapter 2 "Offensive Information Operations", Sun Tzu's proverb on deception is cited to support the discussion of military deception operations. In chapter 5 "Information Operations Planning", Sun Tzu's maxim of "know yourself and know the enemy" is noted down to provide a basis to discuss the importance of knowing the adversaries strategic and operational centers of gravity and guidance to defeating them in the operation planning information (US Joint Chiefs of Staff 1998, V-3).

In 2001, the second edition of British Defense Doctrine (BDD), which sits at the pinnacle of the UK's hierarchy of joint doctrine publications, conveys a message about the tone and nature of the British approach to military activity at all levels. It states that origins of the principles of war can "be traced back to Sun Tzu, they were inherent in Clausewitz's writing, they were first promulgated within the British Armed Forces in the inter-war years, and they achieved their current form under Montgomery's direction immediately after the Second World War" (The Joint Doctrine and Concept Center UK 2001).

5.3.2 *Sun Tzu in military educational institutions*

Our investigation into Corpus 2 also reveals that the translations of Sun Tzu have become a significant presence and have been acquiring increasing interest in Western military educational institutions since the 1980s.

According to a survey conducted by Johnston (1999), the interest of the US Professional Military Education (PME) system in Sun Tzu's strategic thinking has been somewhat constant since the 1980s and 1990s. The translated Sun Tzu emerged in multiple courses on the history of strategic thought or on Asian military thought and practice, as well as in sections of core courses on strategic theory. Sun Tzu is usually introduced as a comparison with Clausewitz. Generally, the similarities between Clausewitz and Sun Tzu are emphasized more than their disparities. Among the US PME institutions teaching Sun Tzu are the National Defense University, Army War College, The US Military Academy (WestPoint), Air University, Naval War College, and US Marine Corps University. This means that most officers attending major US PEM institutions are at some point exposed to at least some part of the translated Sun Tzu text.

Since the beginning of the twenty first century, Sun Tzu has gained increasing attention in US military education institutions and has been taught with more frequency. It has entered the Joint Professional Military Education (JPME) system which aims to strengthen combined and joint operations of military services. JPME courses are available at five levels at a number of

PME colleges and Institutions. On October 6, 2009, the Institute for National Strategic Studies (INSS) and National War College jointly hosted a conference on "Sun Zi's *Art of War* and US Joint Professional Military Education" (Saunders 2009). This was the first conference to bring together leading experts and faculty from major US military academies and senior service schools to address a range of topics on how to teach Sun Tzu, including identifying best practices, translation issues and research gaps. This conference fulfilled three objectives. The first objective was to gain an understanding of how *AOW* is currently taught in US military academies and senior service schools. It found that Sun Tzu was the second most widely taught military strategist, following Carl von Clausewitz and serving primarily as a counterpoint to Clausewitz. "Of the eight institutions represented at the conference, five include *AOW* as part of the core curriculum, while the other three introduce *AOW* in elective courses. The Naval War College and Air Force Academy offer elective courses dedicated specifically to Sun Tzu" (Saunders 2009, 1). Some institutions require students to master and apply its core ideas into various contexts.

The second objective was to discuss how *AOW* fits into a broader historical and cultural context. Participants agreed that Sun Tzu provides good insight into analyzing a number of contemporary and historical strategic issues. Principles of Sun Tzu can be used as "a tool for strategic analysis", "a potential source of ideas for US military operations", "a way to understand the strategic thinking of Asian allies and facilitate cooperation", and "a way to understand the thinking of a potential adversary" (Saunders 2009, 1). Participants also gave brief introductions to which translations are used in courses and why. They reflected on the merits and demerits of various translations of Sun Tzu, highlighting the challenges involved in translation and concluded that the widely used Samuel Griffith translation has a number of advantages for teaching purposes. The third objective involved identifying the best practices in teaching Sun Tzu, recommending future research to inform JPME curricula, and identifying opportunities to engage with international counterparts. For instance, it is recommended that PME faculty who are not experts on Chinese military culture should read Arthur Waldron's "Foreword" and "the Translator's Introduction" in Victor Mair's translation to gain a bird's eye view of the nature of *AOW* and its historical context, an explanation of key concepts and a clarification of how translation choices affect understanding.

It is rather evident that major US military education institutions in different service branches have become increasingly active in teaching Sun Tzu with aid from its translations. To further understand the teaching of Sun Tzu in Western PME, especially in the US, an investigation is conducted on several universities in the following aspects: reading lists for the students, courses in their curriculum and theses finished by master or PhD program students.

National Defense University

Sun Tzu's text was introduced formally into the curriculum of the National Defense University in 1984, when the Department of Military Strategy at the National War College published a book entitled *The Art and Practice of Military Strategy* (Thibault 1984) as the core text on strategy for senior military officers. The textbook is composed of three parts: thinking about strategy, fundamental strategic concepts and the practice of strategy. Sun Tzu was introduced in part two, along with Clausewitz, Mahan, Hart and Corbett. The first chapter of Griffith's translation was copied here. In the introduction to the excerpt, it is stated that for Sun Tzu and Liddell Hart, the military objective was the enemy's will and Sun Tzu preferred the indirect approach to the direct attack on enemy forces. Since the mid-1980s, Sun Tzu has been taught in at least one of the core courses on strategy at the university. For instance, in the academic year 1999–2000, Course 5602 "Fundamentals of Military Thought and Strategy" explored the evolution of the most influential and important thoughts about war in the modern era. Among 22 topics in the course, "Topic 10: The Strategic Dilemma Resolved? The Indirect Approach" narrated that Liddell Hart, horrified by the destruction of the First World War, sought an alternative to reduce the inherent loss of life and property. Hart found his answer in Sun Tzu, who stressed on the importance of *cheng* and *ch'i*, declaring that "the normal (*cheng*) force fixes or distracts the enemy; the extraordinary (*ch'i*) forces act when and where their blows are not anticipated". Sun Tzu's strategic thinking is introduced in comparison with the indirect approach of Liddell Hart and Griffith's translation is a required reading book.

According to our survey, the teaching of Sun Tzu at the National Defense University produced three course papers and three research reports quoting the translated Sun Tzu. One course paper entitled "Fight Against Terrorism: Sun Tzu Revisited" quotes from Griffith's translation Sun Tzu's strategic principles such as foreknowledge and information, deception and victory without fighting to discuss the war against terrorism (Roberson 2002).

In addition to the courses, the Information Resources and Management College at the National Defense University has held an annual essay competition called the "Sun Tzu and Information Warfare" since the mid-1990s. This competition is designed to inspire research on definitions of information warfare, the implications for doctrine and operations, the implications for organizational relationships between government and private industry and the vulnerability of national information structures, among other topics. In 1997, a book including a set of winning essays entitled *Sun Tzu and Information Warfare* was published (Neilson 1997).

Army War College

In the branch of the US Army service, Sun Tzu is often listed in the Army Chief of Staff's professional reading list, and in 2014 it was highly prioritized in the US Army Intelligence Center of Excellence Commanding General's reading list.

At the Army War College, Sun Tzu is taught in at least three courses: an advanced course on the Chinese military, a course on regional security in

Asia and a required course on strategic theory. The first course cites Ames' translation; while the latter two Griffith's. The course "The Theory of War and Strategy" recommends students to read Michael Handel's comparison of Clausewitz, Jomini and Sun Tzu (Johnston 1999, 18). It begins with a study of Clausewitz, then moves on to a lesson on "Sun Tzu, Mao and Asian Military Thought" (which is about three hours of instruction). The lesson focuses on fundamental concepts such as the indirect approach, and the principle of deception as well as winning without war. Based on Michael Handel's comparative analysis, the differences between Clausewitz and Sun Tzu are stressed. In addition, the lesson compares Sun Tzu and Mao's views on guerilla warfare.

The instruction of Sun Tzu at the Army War College is rather productive in spreading the strategic theory in *AOW*. According to our survey, 30 monographs, book chapters and research reports with quotes from the translated Sun Tzu were published by the Army War College and authored by its faculty members as well as students. In addition, there were three master degree theses headed with Sun Tzu: "Sun Tzu: Ancient Theories for a Strategy against Islamic Extremism" (Rice 2006), "Sun Tzu: Theorist for the Twenty First Century" (Wilcoxon 2010), and "21st Century Strategy Needs Sun Tzu" (Critzer 2012), all focusing on the Sun Tzu's relevance to modern strategic issues.

The US Military Academy (West Point)

In 2012, the US Military Academy (West Point) selected *AOW* (both Giles' and Griffith's translations) for its Top 10 Military Classics. In 2014, Sun Tzu was again included in The Officer's Professional Reading Guide Top 100 List.

At West Point, there are two courses taught about Sun Tzu, one is "HI385 War and Its Theorists" in the Department of History, the other is "SS493 Studies in Grand Strategy" in the Department of Social Sciences. The course "War and its Theorists" discusses commanders who theorized about the nature and conduct of war, the relationship between politics and strategy, and the impact of warfare upon society. The course uses Sawyer's translation and treats Sun Tzu as one of the best military theorists around the globe alongside Clausewitz, Jomini, Mahan, Fuller and Hart. The other course "Studies in Grand Strategy" examines the theory and practice of grand strategy diachronically in order to identify lessons and transferable principles to guide the formulation of future US policy. Sun Tzu is treated as one of the grand strategic theorists like Thucydides, Machiavelli and Clausewitz.

Air University

At the Air War College of Air University, Sun Tzu's military theory is taught through a core course in the Department of Strategy, "Doctrine and Airpower" (Johnston 1999). The first section of the course examines theorists of war, including Sun Tzu, Clausewitz, Jomini, Mahan and Corbett. Also, in the School of Advanced Airpower Studies, a course is offered on the foundation of military theory, beginning with Sun Tzu, and moving chronologically to Clausewitz, Jomini, Du Picq, Fuller, Hart, Tukhachevski, Mahan, Corbett and Mao. Sawyer's translation is used in this course. One seminar is devoted to Sun Tzu in comparison with Clausewitz. In the Department of Warfighting, Professor George J. Stein offers an elective graduate course entitled: "*The Art of War* Sun Tzu, The Seven Military Classics and Unconventional Strategic Thought" with 13 to 15 sessions (Stein 2007). This seminar discusses strategic thinking primarily by inquiring into several major works of classical Chinese strategy. Students are required to analyze, with some depth, the relationship between the approach of Sun Tzu's *Art of War* and contemporary strategic thought. To better understand chapters of *AOW*, four translations are recommended: Griffith, Ames, Sawyer and Denma.

The lessons taught in Air University have triggered insights into the application of Sun Tzu in modern context. At least seven papers authored by students and faculty members include "Sun Tzu" in their titles, mimic the title of *The Art of War,* or use quote from *The Art of War* in their title. The following examples indicate the enthusiasm of using Sun Tzu: "Sun Tzu in the Age of Technology" (Johnson 1992), "The Gettysburg Campaign: 'Know Your Enemy, Know Yourself'" (Mendez 1995), as well as "Sub Tzu and the Art of Submarine Warfare" (Borik 1995).

Naval War College

The US Naval War College consists of four colleges (College of Naval Warfare, College of Naval Command and Staff, College of Distance Education, College of Operational and Strategic Leadership) and other centers. Since the 1980s and 1990s Sun Tzu has been taught "as part of a core required course on strategy" (Johnston 1999, 17). In 2006, Naval Command and Staff College offered a 57-hour core course at senior level "Strategy and Policy", in which one topic "Strategic Concepts in War Fighting" takes nine hours. This topic covers Clausewitz's and Sun Tzu's land strategy; Mahan's and Corbett's naval strategy; and Douhet's, Mitchell's and Boyd's air strategy. It can be seen that Sun Tzu's share in the course is rather small. However, great changes took place in 2008, when the same course developed in synchronization with JPME Objectives was taught at both the College of Naval Warfare and the College of Naval Command and Staff. There are 13 case studies designed for this course, among which the first one is "Masters of War: Clausewitz, Sun Tzu, and the Development of Contemporary Strategic Thought", where Sun Tzu and Clausewitz are compared in great details. Griffith's translation is listed as required reading. Specifically, 17 questions are raised about Sun Tzu in ten case studies, while 19 questions are raised about Clausewitz. The questions about Sun Tzu cover important concepts in *AOW*, such as: (1) Evaluate the role of intelligence in *The Art of War.* Would Clausewitz agree with the Sun

Tzu view? Which view is more relevant today? (2) Which theorist—Clausewitz, Sun Tzu or Mao—provides the best insight into why Communist strategy in Vietnam was successful? (3) Which theorist do you regard as more relevant to the current global war on terrorism, Clausewitz or Sun Tzu? These well-designed questions help to develop an informed modern understanding of Sun Tzu. In the academic year 2015–2016, the course "Strategy and Policy" has been accredited as a Joint Professional Military Education Phase II senior level course. Among the 13 case studies of the course, 11 contain 37questions on Sun Tzu; however, only 10 include 30 questions on to Clausewitz. It is evident that *AOW* and *On War* constitute the cornerstone of this course, and Sun Tzu began to gain increasing significance (Lee 2010, 120). One of the faculty members of Naval War College, reflected that teachers seek "to use this old Chinese text to generate new ideas". For instance, they turned Sun Tzu's "know your enemy and know yourself" principle into an elaborate process of net assessment, used Sun Tzu's attacking strategy maxim to deliberate types of conflicts including not only The Cold War, but also a Maoist insurgency or of Al Qaeda and Associated Movements (AQAM), and pondered the notion of hybrid warfare through Sun Tzu's discussion of *cheng* and *ch'i* forces (Lee 2010, 120).

Another core course named "Strategy and War" also teaches *AOW*. It was labeled as a Joint Professional Military Education Phase I Intermediate Level Course and offered in the academic year 2015–2016, it included 12 case studies. The course begins with "Case Study 1 Masters of War Clausewitz, Sun Tzu and Mao". Griffith's translation is listed among the required readings and Sun Tzu's section involves seven case studies with 17 questions altogether. These questions are set in both historical and contemporary strategic context, such as (1) Which theorist, Sun Tzu or Clausewitz, best explains the outcome of the Korean War? (2) Sun Tzu advised that the best way to win is to attack the enemy's strategy. How, and to what extent, does that insight apply to the war between Al Qaeda and Associated Movements, and the American-led alliance? These questions have shown that the translated Sun Tzu is studied as a significant reference to modern strategic issues.

In addition to the core courses, two elective ones concerning Sun Tzu were also offered at Naval War College in the academic year 2009–2010: "EL 531 Sun Tzu's *The Art of War*" and "EL 637C Applying Analysis to Warfare: Sun Tzu with a Calculator". EL 531 is probably the first course exclusively devoted to Sun Tzu in the US PME system. It was composed of three sections. The first section introduces the historical significance of *The Art of War*, its relevance for contemporary strategy, and its relation to a universal military theory and to Asian approaches to strategy. The second section deliberates on the text itself, the style of writing, commentaries on the text, the internal coherence, logic, etc. The third section examines the historical context of the text: politics, strategy, tactics and technology of warfare from the Chou Dynasty to the Warring States period. According to Professor A. Wilson, this lesson aimed to enhance the appreciation of the strengths and weaknesses of Sun Tzu approach to strategy and operations with critical analysis. It involved a close reading of several translations of Sun Tzu, including Sawyer's and Ames'. The other course, EL 637C, by drawing upon Sun Tzu's concepts, surveys techniques in measuring risk and uncertainty at the operational and tactical level of war.

US Marine Corps University

It is most likely that among all the professional military universities discussed, US Marine Corps University is the institution where Sun Tzu has played the greatest role in the curriculum. The US Marine Corps University is a professional military education institution set up in 1989 by Alfred Gray, then Commandant of the Marine Corps. It consists of the Marine Corps War College, Marine Corps Command and Staff College, School of Advanced Warfighting, Expeditionary Warfare School (formerly Amphibious Warfare School), School of Marine Air-Ground Task Force Logistics and others.

AOW becomes the 1990 "Commandant's Choice" for the Professional Military Education (PME) reading program. The "Commandant's Choice" is a book or a group of books chosen once a year for their timeliness, value and Corps-wide interest. General Alfred Gray explained that "*The Art of War* is among the greatest classics of military literature ever written" and further urged all marines to read the book, "thinking about its meaning, and using it to initiate a Corps-wide professional discussion on the art of war" even though Sun Tzu is already assigned for the ranks of first sergeant, master sergeant, chief warrant officer 3 and captain (Anonymous 1990). Till present, Sun Tzu remains in the Marine Corps Professional Reading List, which includes both Griffith's and Cleary's translations.

At the Marine Corps War College, Sun Tzu's *AOW* is learned in a required course on "War, Policy and Strategy", in which Sun Tzu is compared with Clausewitz through the lesson "The Relevance of Sun Tzu and Clausewitz to the Political, Strategic, and Operational Levels Of War" (Candela 1998, 299). This lesson requires students to analyze the theories of war proposed by Sun Tzu and Clausewitz, and their relevance to military effectiveness at the political, strategic and operational levels of war. One of the specific objectives is to compare Sun Tzu and Clausewitz to *FMFM 1 Warfighting* and *FMFM 1– 1, Campaigning*. Topics for discussion concerning Sun Tzu include attacking enemy's strategy; general's autonomy of command on the battlefield, the use of an army's strategic advantage, and extraordinary forces. Yuan's translation of *AOW* is used together with Handel's comparison of Sun Tzu and Clausewitz and Griffith's translation is recommended for further reading.

In the Command and Staff College, Sun Tzu is also taught in the Course 8701 "Theory and Nature of Warfare" for the nonresident program. According to its 1993 syllabus, this course offers a 68-hour study of warfare from the age of limited war to the present in 13 lessons. It opens with "Lesson 1 Sun Tzu" which lasts six hours. The objectives of this lesson are to explain Sun Tzu's view on the nature of policy and war and to examine Sun Tzu's theories and his approach to the development and purpose of a theory of war. In this lesson, students are required to read Griffith's translation of Sun Tzu. Questions are asked with specific attention to the relationship between politics and war, the validity of Sun Tzu in conditions of modern high-tech warfare, the relationship between Sun Tzu and contemporary warfare (such as Operation Desert Shield Storm). This course is also available to the residents program at the Command and Staff College. However, the educational objectives for this lesson are slightly different: (1) to place Sun Tzu into his historical context, (2) to evaluate Sun Tzu's basic theories and his approach to warfighting and (3) to appreciate the manner in which *FMFM1 Warfighting* incorporates Sun Tzu's concepts (Candela 1998, 110). Students are required to attend a lecture on Sun Tzu given by Dr. Arthur Waldron, a faculty member at the US Naval War College, and to read Griffith's translation. Questions are also asked about Sun Tzu, one of which concerns the principle of bloodless victory.

In addition to the above PME institutions, Sun Tzu also appear in the courses provided by other US universities: "Strategy and Policy" at the University of California, Los Angeles (2001), "War, Peace and Strategy" at Columbia University (2011), "Studies in Grand Strategy" at Yale University (2013), "Chinese Grand Strategy: Foreign and Military Policy" at the Institute of World Politics Press (2014) and "Strategy, Policy, and War" at the University of Pennsylvania (2014).

AOW has not been taught only in US colleges and universities, it also finds its way into military higher learning in other Western countries, like the UK, Australia, Canada, Netherland, etc. For instance, Sun Tzu is taught in the course of "Strategy and War" at the Southampton University (UK), "Intelligence and Security" at Stanfordshire University (UK), "War and Security" at Ottawa University (Canada), "Strategic Studies" at the Australian National University, "Graduate Seminar on Military Theory" at the Baltic Defence University (Estonia, Latvia and Lithuania).

Sun Tzu has also appeared in the Royal Australian Navy Reading List (2006), The Australian Chief of Army's Reading List (2012) and The Canadian Army Reading List (2009). According to the survey conducted by McElhatton (2014), Sun Tzu appears 23 times in the global Recommended Leadership Texts of Professional Military Reading Lists, 13 times in the US and 10 elsewhere. *AOW* ranks as the top ancient text most frequently listed, followed by Thucydides' *History of the Peloponnesian War* and Caesar's *Commentaries*.

5.3.3 *Sun Tzu on battlefields*

The translated Sun Tzu appears not only on textbook pages, but also on the operation blueprints for battles and wars. Since the 1990s, the US military began to implement what they have gained from Sun Tzu into a series of wars: the Gulf War, Iraq War, Afghanistan War, etc.

The Gulf War

The Gulf War was an international conflict fought between Iraq and a US-led coalition, which was triggered by Iraq's invasion of Kuwait in August 1990. The war began with the allied coalition's military offensive against Iraq on January 17, 1991. The massive US-led air bombardment, namely Operation Desert Storm, lasted throughout the war and destroyed Iraq's air defenses, communications networks, government buildings, weapons plants and oil refineries. After mid-February the allies shifted their air attacks to a massive ground offensive named Operation Desert Sabre against Iraq's ground forces in Kuwait and southern Iraq. Shortly later, Iraqi resistance completely collapsed, and US President George Bush declared a cease-fire on February 28. The Gulf War ended with a decisive victory for the US-led coalition at a small cost of 300 lives against the estimated Iraqi military deaths ranging from 8,000 to 100,000.

It is not surprising that the US-led coalition forces would win the Gulf War. However, people may wonder why the allies, greatly outnumbered by the Iraqi troops, prevailed in such a short period of time with such ease and few casualties. In addition to the intense use of high-technology weapons, the right choice of strategy is the key factor that led to the victory. The Gulf campaign as a whole followed the Sun Tzu model, displaying Sun Tzu's teachings "to the fullest and truest extent" (Marshall 1999, 58).

Firstly, the masterminds of the Gulf War are disciples of Sun Tzu, such as Schwarzkopf, Boyd, Warden and Franks. General Norman Schwarzkopf, who led the US and coalition forces in the Gulf War, was "a student of Sun Tzu and employed tactics from *The Art of War* to secure victory" (Phua 2007, 47). John Boyd was "the chief architect behind the application of Sun Tzu in the Gulf War" (Phua 2007, 47). Boyd had a direct role in the planning of Operation Desert Storm. According to Coram (2002), the Secretary of Defense, Dick Cheney, invited Boyd to the Pentagon to discuss strategy. With Boyd as his background adviser and the support from Colin Powell (then Chairman of the Joint Chiefs of Staff), Cheney favored a "left-hook" maneuver led by the marines. John Warden, a retired US Air Force colonel, is best known for drawing up the 1991 Gulf War air campaign. He was "a strong proponent of psychological effects in war, the 'indirect' approach of Sun Tzu and Liddell Hart", advocating that planners focus less on destroying and killing and more on how to create chaos, confusion and paralysis (Sloan 2012, 58). Brigadier General Tommy Franks, a graduate of the Armed Forces Staff College and Army War College, was reported "to be a devotee of Sun Tzu and often found quoting him" (Phua2007). He served as Assistant Division Commander (Maneuver), 1st Cavalry Division, during the 1991 attack into southern Iraq.

According to a news report from the Los Angeles Times on February 18, 1991, even prior to the ending of the Gulf War, Col. Michael D. Wyly, vice president of the Marine Corps University, predicted that Sun Tzu's teachings would have a noticeable impact in the Persian Gulf land battles, "particularly in the tactics used by the younger officers who have been heavily exposed to Sun Tzu's military classic" (Richter 1991). Other analysts expected the marines to follow Sun Tzu's approach to avoid a frontal assault, and use landing craft, helicopters and Harrier vertical takeoff jets to punch through vulnerable spots in the Iraqi lines, then turn to strike them from the rear (Richter 1991).

Secondly, Sun Tzu's strategic thinking rings loud and true in the three phases of the Gulf War. As Marshall (1999, 57) noted, in Sun Tzu's view, there are three phases to all successful wars: breaking the enemy's strategy, breaking the enemy's alliances and fighting. In the first phase of the Gulf War, coalition forces contained the Iraqi forces in Kuwait through the mobilization of a rapid reaction force to Saudi Arabia. In the second phase, the diplomatic efforts through the UN Security Council and through negotiations with Iraq's Arab neighbors deprived Iraq of political, economic and military support. In the third phase, as Sun Tzu advocated, the Gulf War featured maneuver, deception, intelligence gathering, surprise and psychological operations. General Schwarzkopf's "Hail Mary" maneuver, was a classic example of Sun Tzu's teachings. He argued that a commander should: "Go into emptiness, strike voids, bypass what he defends, hit him where he does not expect you" (Marshall 1999, 58).

Specifically, at the level of tactical and operational strategy, the Gulf War was "a conventional war and fighting it using Sun Tzu's indirect approach was sufficient to deliver an overwhelming victory, which Schwarzkopf did" (Phua 2007, 46). Saddam Hussein's strategy in the war was a direct frontal assault based on fortified positions. His entrenched Iraqi forces were waiting for an allied head-on attack from south (Romm 1991). However, the allied planned the war in a different manner. Abandoning the traditional, muddling-through warfare by attrition approach, "the US fought smarter for a change in the Gulf War" (Romm 1991). The allied troops choose to flank the Iraqis in the "left-hook" maneuver. And they also did everything possible to deceive Hussein into expecting an amphibious frontal attack from the east by the allied troops. The 1st Cavalry Division conducted feinted frontal raids against the major trench and obstacle works of Iraq troops, using the *cheng* method. Meanwhile, other US Corps maneuvered to the west to envelop Iraqi front-line divisions and destroy the Republican Guard's divisions further to the north, using the *ch'i* method. In an interview shortly after the Gulf War, Tommy Franks cited Sun Tzu as having inspired the effective deception plan. "Sun Tzu said 2500 years ago, 'Make your way by unexpected routes and attack unguarded spots.' That's what the campaign plan called for, and that's exactly what we did … it was a masterstroke" (Franks and Hollis 1991). The operation became "a first rate instance of Sun Tzu's *cheng/chi*" tactic with the surrender of fifteen Iraqi divisions to two US Marine divisions (Phua 2007, 46).

To know the enemy and the self, the coalition obtained and exploited critical information about their adversary they gained from aircrafts and satellites far before the war began. Thus, "Operation Desert Storm legitimized information warfare as a warfighting strategy" (Gillam 1997). When bombing the Iraqi Republican Guard, allied airplanes also dropped leaflets to persuade the adversary troops into surrender. This psychological operation worked so well that most of the 87, 000 Iraqi prisoners taken during the war entered captivity with allied leaflets in hand (Marshall 1999).

To sum up, the Gulf War proves "how effective Sun Tzu's theories could be in a modern technology-led war" (Marshall 1999, 60). It shows that *AOW* is as valid now as it has been 2500 years ago, even to the point of directing the actions of Western military superpowers. "You could say Sun Tzu's spirit is hovering above the whole conflict", remarked Col. Sam Gardiner, a retired Air Force officer (Richter 1991).

The Iraq War

In addition to the Gulf War, the US-led coalition fought the Iraq War once again under the guidance of the Chinese strategist Sun Tzu.

The Iraq War, also called Second Persian Gulf War, lasted eight years from 2003 to 2011. It consisted of two phrases. The first phrase, also referred to as Operation Iraqi Freedom, was a brief, conventionally fought war from March 19 to May 1, 2003, when the coalition troops of the US and the UK invaded Iraq and rapidly defeated the Iraqi military and paramilitary forces. The second phase lasted a rather long period, with the US-led occupation of Iraq being opposed by insurgents. December 2011, the US military formally declared the end of its mission in Iraq.

Sun Tzu's influence on the war can be detected from the following facts. Firstly, the masterminds of the Iraq campaign, former Secretary of Defense Donald Rumsfeld and General Tommy Franks were frequently found quoting Sun Tzu (Holmes 2003). Franks once replied, when asked to estimate enemy forces, "as has been the case since Sun Tzu said it, precise knowledge of self and precise knowledge of the threat leads to victory" (Hyde 2003 March 29). Before the air war against Iraq began, Brigadier General Russell Sutton, director of the Marine Corps operations division, was reported as saying, "[t]he last thing we want to do is to meet him [Hussein] head -on. This is pure Sun Tzu" (Romm 1991, 160).

Secondly, in December 2002, the Pentagon distributed four books to service members overseas or those aboard ships at sea: *AOW, Medal of Honor, War Letters: Extraordinary Correspondence from American Wars* and *Henry V* (Rhem 2002). This edition of *AOW* was Giles' translation published by Dover Publication. The widespread and intensive reading of *AOW* among American soldiers has laid a solid foundation for Sun Tzu's impact on the war.

More importantly, another book inspired by Sun Tzu's concepts, *Shock and Awe: Achieving Rapid Dominance,* became the official handbook to this conflict according to a Pentagon document (Hyde 2003 March 29). *Shock and Awe* (Ullman and Wade 1996) was prepared by Defense Group Inc. for the National Defense University. It proposes a military theory based on the use of overwhelming power and spectacular displays of force to destroy the enemy's will to fight. According to Secretary of Defense Donald Rumsfeld, Shock and Awe is "a way to get people to do what we wanted and stop doing things that we did not want—or to win the war without having to fight the battle" (Ullman 2010, 80–81). This reveals that the doctrine of Shock and Awe actually follows Sun Tzu's principles. In the book, Ullman and Wade remarked that:

> Sun Tzu was well aware of the crucial importance of achieving "shock and awe" prior to, during, and in ending the battle. He also observed that "war is deception," implying that Shock and Awe were greatly leveraged through clever, if not brilliant, employment of force.
>
> (Ullman and Wade 1996, 19–20)

They also listed four characteristics of the Shock and Awe principle: (1) absolute knowledge and understanding of self, adversary and environment; (2) rapidity and timeliness in application; (3) operational brilliance in execution; and (4) (near) total control and signature management of the entire operational environment (Ullman and Wade 1996, xii). Both the first and second are consistent with Sun Tzu's teaching.

To illustrate the concept of Shock and Awe, nine historical examples of military applications that fall within this concept are listed. Among them, the fifth case is the story of Sun Tzu's training of the court concubines and the decapitation of two of them (Ullman and Wade 1996, 27–29). The story is retold to illustrate the way to achieve Shock and Awe: with selective, ruthless and rapid application of force, we can harm the few principal targets and convince the majority that resistance is futile. It is evident that the term "decapitation", repeatedly used by America in their efforts to kill Saddam, derives from Sun Tzu.

The Pentagon used the Shock and Awe strategy to the full extent in the Iraq War. Over 30,000 bombs and missiles were dropped on Baghdad within the first 48 hours after the war began. Donald Rumsfeld also stressed on "decapitating" the enemy leadership, especially Saddam Hussein. In April 2003, Iraqi forces collapsed, Saddam fled and his regime was pulled down. Shock and Awe, followed by "decapitation", helped the US win the major battles in Iraq in less than a month (Hwang and Ling 2009).

Thirdly, Sun Tzu's guidance can also be detected in many other aspects of the war. For instance, the extensive psychological operations that took place during the war and led to the mass surrenders of Iraqi troops are also consistent with Sun Tzu's principles (Morgan 2005, 102). The quick tempo of coalition operations reflected the advice of *AOW* (Morgan 2005, 103). As Sun Tzu advocated taking forces and territory without destroying resources and avoiding prolonged operations, Operation Iraqi Freedom, using precision weapons extensively, was "one of the most conscientious about preserving enemy non-combatants and avoiding destruction to the nation's resources"

(Morgan 2005, 101). Sun Tzu's approach of *cheng* and *ch'i* was once again employed in the battlefield. In a real sense, the US used air power in the *cheng* role, pummeling Iraqi ground units while allowing the 3rd Infantry Division and the 1st Marines to function in the capacity of *ch'i*. The speed with which coalition ground forces advanced and their readiness to bypass urban areas had thrown the Iraqis off balance (Holmes 2003). In addition, referring to Sun Tzu's principle of deception, US forces took efforts to "trick, cajole, induce or force the adversary" (Ullman and Wade 1996, 29).

On a whole, as Holmes (2003), Macintyre (2003 April 12), Morgan (2005), McNeilly and McNeilly (2001) and many other analysts believe, the successful operations before and during Operation Iraqi Freedom are consistent with Sun Tzu's strategic paradigm.

In addition to the Gulf War and Iraq War, Sun Tzu became the guru behind the US invasion of Afghanistan starting from the year 2001, which was led by US General Tommy Franks. In 2001, *AOW* was the ninth most ordered book by marines (Hyde 2003 March 29). McNeilly and McNeilly (2001, 4) believed that "Sun Tzu's philosophy was again demonstrated by American general Tommy Franks", who was fond of quoting from *AOW* and used Sun Tzu's philosophy to drive the Taliban from power in a few short months. Zweibelson (2010) agreed that the US forces in Afghanistan follow Sun Tzu rather than Clausewitz in their operations.

To sum up, the influence of the translated Sun Tzu on these US-led wars is exhibited in at least three aspects. Firstly, not only the masterminds and commanders of the military conflicts were advocates of Sun Tzu, but also the soldiers fighting the wars were readers of the translated *Art of War*. Secondly, Sun Tzu's strategic wisdom was used not only during the war at the operational or tactical level, but also before the war at the grand strategy level. Thirdly, Sun Tzu's principles were used in many aspects, in combination with the new advancements in technology and the innovation of military theory attuned to a specific context, such as Shock and Awe.

The successes of these contemporary wars, as well as those in history, have convinced people in the military sphere of the validity of Sun Tzu. Such conviction further pushes forward Sun Tzu to non-military competitive fields. This suggests an inevitable trend of Sun Tzu's expansion from military institutions into a far wider area of society.

5.4 At the societal level

The previous section found that Sun Tzu is firmly rooted in the institutional discourse of US strategic culture. This integration between Eastern and Western strategic culture has gradually expanded beyond the military to a wider range of Western discourses: business, management, sports, pop culture, etc., as shown in Corpus 3. Although the translated Sun Tzu in the non-military sphere is not the focus of this study, it is worthwhile to be investigated briefly so as to provide a better understanding of how *AOW* impacts Western culture as a whole. Moreover, it is believed that such expansion will further ensure the canonical status of the translated Sun Tzu at the societal level.

A survey of the translations quoted in Corpus 3 finds out Giles' translation has been quoted 81 times, Griffith's 193 times, Sawyer's 183, Ames' 172 and Cleary's 140, which suggests that a text in Corpus 3 quotes the translated Sun Tzu more than nine times on average, even higher than the average quotes in Corpus 2.

5.4.1 Sun Tzu in business and management

In addition to its application in military discourse, *AOW* is tremendously popular in the competitive arena of business and management. Sun Tzu's translations have been quoted lavishly in the Western discourse on business and management. A large number of books apply Sun Tzu's principles in business branding, strategy, marketing, intelligence, management, manufacturing and personnel. Successful business leaders, from the board room to the shop floor, have employed Sun Tzu. Business planners and researchers refer to Sun Tzu as a master for strategic inspirations.

A typical example is *Sun Tzu and the Art of Business: Six Strategic Principles for Managers,* authored by McNeilly (1996, 4–5), who articulated that "business by definition deals with competition, Sun Tzu's principles are ideally suited to competitive business situations". Professor McNeilly has served as a global marketing executive and has several years of experience with both IBM and Lenovo in the IT industry. He distilled from Sun Tzu's military theory the six most significant principles and discussed them in six chapters respectively, drawing quotations from Griffith's translation and expounding with application cases from the business battlefield. The six principles are: (1) win all without fighting: capturing your market without destroying it; (2) avoid strength, attack weakness: striking where they least expect it; (3) deception and foreknowledge: maximizing the power of market information; (4) speed and preparation: moving swiftly to overcome your competitors; (5) shape your opponent: employing strategy to master the competition; and (6) character-based leadership: providing effective leadership in turbulent times. For the first principle, McNeilly quoted 18 times from Griffith's translation of *AOW*, among which four quotes are from the Chinese paratexts, specifically the translated comments. The second principle is discussed in chapter 2, which begins with the direct quotation of the "water simile". It is argued that for companies, head-on attacks against each other and battles of attrition are very costly and time consuming, leaving both sides in a weakened state. Instead, using the method of avoiding strength and attacking weakness will minimize the use of resources, maximize the gains and eventually increase profits.

Other similar book titles applying Sun Tzu to business and management include: *The Art of War for Executives* (Krause 1995), *Sun Tzu: The Art of War for Managers: 50 Strategic Rules* (Michaelson 1998), *Sun Tzu's The*

Art of War Plus The Art of Marketing (Gagliardi 2003), *Doing Business in China: The Sun Tzu Way* (Brahm 2004), *Sun Tzu Strategies for Selling: How to Use the Art of War to Build Lifelong Customer Relationships* (Michaelson and Michaelson 2003a), *Sun Tzu Strategies for Marketing:12 Essential Principles for Winning the War for Customers* (Michaelson and Michaelson 2004), *The Art of Business in the Footsteps of Giants* (Yeh and Yeh 2004), *Sun Tzu and the Project Battleground: Creating Project Strategy from The Art of War.* (Hawkins and Rajagopal 2005), *Striptease: The Art of Corporate Warfare* (Deva 2005), and *The Art of War for Small Business: Defeat the Competition and Dominate the Market with the Masterful Strategies of Sun Tzu* (Sheetz-Runkle 2014). It seems that there is an endless list of such books as new titles keep coming out. They have been putting forward advice for business managers, quoting intensively and extensively from the translated Sun Tzu.

Sun Tzu is not only quoted in books and papers but also practiced in business and management. According to Wheeler (2014), scores of chief executives surveyed by Forbes cited *AOW* as their favorite business book. Larry Ellison, founder and CEO of Oracle, is an admirer of *AOW* and a master of applying Sun Tzu's battle tactics to the modern-day warfare of business competition. Ellison followed Sun Tzu's tenant that a smaller force can beat a larger one by causing its rival to respond before thinking. This is the way Oracle competed against SAP. Salesforce.com CEO Marc Benioff remarked in an interview: "Larry consistently executes *AOW* better than any CEO. SAP never should have reacted to Oracle's statements because it makes customers and investors view Oracle as a peer to SAP, when they aren't" (Garner 2006). Larry also took to heart Sun Tzu's advice of selecting people and giving them responsibilities commensurate with their abilities when he examined Oracle's channel shortcomings and decided to hire former Hewlett-Packard CEO Mark Hurd as president (Burke 2011).

Michael Ovitz, the top agent who allegedly controls Hollywood, attributes his success to the teachings of Sun Tzu rather than modern management gurus (Nisse 1999; Zeman 2011). Ovitz was the co-founder and chairman of Creative Artists Agency until 1995, president of the Walt Disney Company from 1995 to 1997, and the founder of the Artists' Management Group in 1999. As one of the best-known exponents of Sun Tzu, Ovitz begins each day reading tracks of *AOW* (Hall 2002). He also handed his senior executives copies of *AOW* when he was running the Creative Artists Agency (Nisse 1999). For him, Sun Tzu's aphorisms were very relevant to the new company he has set up.

In 1996, a company named "Sun Tzu Security" was founded as one of the first providers in US to offer turnkey information security solutions. It aimed to ensure the success of its customers by providing strategic consulting services, such as information security risk assessments and computer investigative services (Anonymous 2000; 2003). Such a company name displays modern manager's appreciation of Sun Tzu's insight into the role of information.

Still many companies are set up to offer training programs in strategic planning, which rely on various translations of Sun Tzu. For instance, The Science of Strategy Institute founded in 1998 by Gary Gagliardi uses his own translation of *AOW*. The Sonshi Group (Sonshi.com) founded in 1999 comes up with a translation by group efforts.

The translated Sun Tzu is also taught in university courses. Corporate raider Asher Edelman made *AOW* required reading for admission to his course on entrepreneurship at Columbia University's Business School (Lawlor 1991), because Sun Tzu preached self-discipline, personal character and the principle of winning wars without combat (Edelman and Edelman 1987). At the Wharton School of the University of Pennsylvania, a course entitled "MGMT 826 East Meets West: Strategic Implications for Managing in the 21st Century" also teaches Sun Tzu's principles in business and management.

5.4.2 *Sun Tzu in sports and entertainment*

The translated Sun Tzu has also been ushered into the discourse of sports in researching, coaching and conducting sports competitions, including sports like football, basketball, golf, etc.

According to Winter (2006), *AOW* has become increasingly popular in football. The Portugal football coach, Luiz Felipe Scolari, had taken inspirations from Sun Tzu to develop successful strategies in football matches (Winter 2006). When he led Brazil to glory at the 2002 World Cup, all of his players had copies of *AOW*. Scolari encouraged his Brazilian team members to absorb Sun Tzu's strategy, saying that "the enemy could be beaten by stealth, speed and above all preparation" (Winter 2006). Scolari's successor, Carlos Alberto Perreira, has continued the book-club tradition and Marcos Evangelista de Morais (better known as Cafu), the team's captain, is also an admirer of *AOW*. When leading the Portugal team and preparing the match between Portugal and England in the 2006 World Cup, Scolari read Sun Tzu's lessons every night (Ipton 2006). By following Sun Tzu's stress on preparedness, he had a well-rehearsed Plan A, a 4–2–3–1 formation, with a rigorous drilling. Like what Sun Tzu said "speed is the essence of war", Scolari believed that the rapid movement of some players, such as Simao and Ronaldo, is a fundamental pillar in the building of victory. Scolari is a revered leader, knowing the art of rewarding and punishing players, according to what Sun Tzu articulated, "too frequent rewards indicate that the general is at the end of his resources; too frequent punishments that he is in acute distress" (Ipton 2006). Other footballers, such as Nigel Reo-Coker, the West Ham United captain, have drawn lessons from Sun Tzu's teachings about the need to "respect and know your enemy" and "anticipate their next move" (Brodkin 2006).

In a journal paper entitled "*Sun Tzu: The Art of War and Basketball*", Ashley (2008) advised how coaches could lead US high school basketball teams to success. His paper, quoting intensively from Griffith's translation, offers suggestions for training, hosting the opposite team, knowing the strategy of yourself and your adversaries, attacking weakness and avoiding strengths, etc. Bob Knight, the famous coach of college basketball, led the Indiana Hoosier team to three NCAA championships and 11 Big Ten titles. He received National Coach of the Year honors four times and Big Ten Coach of the Year honors eight times. According to Wojciechowski (1992), Knight was introduced to Sun Tzu's work in the late 1960s. As a voracious reader of *AOW*, he has adapted many of the teachings to his own coaching. Other basketball coaches, such as Pat Riley, also learned from Sun Tzu for their team building (Harry 1997).

For the game of golf, *AOW* has also become the source of inspiration. Chapin and McDonald (1992) quoted from time-honored truths, tactical maneuvers and noted down the astounding relevance of Sun Tzu to golfers. Wade (2006), in his book entitled *Golf and the Art of War: How the Timeless Strategies of Sun-Tzu can Transform Your Game*, adopted the lessons of warfare in *AOW*. Wade came up with a unique analogy of Sun Tzu's military philosophies and the game of golf. He argued that Sun Tzu's strategy in careful planning, sound information, organization, control and weather elements are the keys to the success of any campaign and can serve as the ultimate attack plan for a golf victory.

Sun Tzu made his first presence in screen when he was cited in a popular movie "Wall Street" in 1989. The movie, directed by Oliver Stone and starring Oscar-winning actor Michael Douglas, tells a story of corporate mergers and hostile takeovers. It highlights Sun Tzu's wisdom in the battle between Gordon Gekko, the film's villain, and Bud Fox, the young hero. Gekko, a "greed-is-good" character, announces his ruthless strategy of financial speculation, "I don't throw darts at a board. I bet on sure things. Read Sun Tzu, *The Art of War*. Every battle is won before it is ever fought. Think about it" (McCann 2012, 27). Later, Bud Fox paraphrases, "Sun Tzu: If your enemy is superior, evade him. If angry, irritate him. If equally matched, fight, and if not split and reevaluate" (McCann 2012, 27). This double quotation of Sun Tzu by two main characters reflected the widespread popularity of Sun Tzu in Western culture, and in turn, it generated another surge of admiration for *AOW* in the US.

The Sopranos, an HBO hit television series about a Mafia family, also quoted Sun Tzu. This program, lasting from 1999 to 2007 with six seasons and 86 episodes, is regarded as one of the greatest television series, has won a multitude of awards and become a staple of 2000s American popular culture. The show attributes the success of the gangster boss and strategist, Tony Soprano, and his criminal organization to Sun Tzu. *AOW* was recommended in Episode 308 by Tony's psychiatrist Dr. Melfi in therapy and quickly spread through the ranks of Tony's fellows. Later, Tony Soprano mused:

> Been reading that– that book you told me about. You know, *Art of War* by Sun Tzu. I mean here's this guy, a Chinese general, wrote this thing 2400 years ago, and most of it still applies today! Balk at the enemy's

power. Force him to reveal himself…But this book is much better about strategy.

Tony also quoted: "If your opponent is of choleric temper, irritate him" (Episode 308). Even his sister began quoting the ancient general, but with an inaccurate pronunciation: "According to San Tee Zoo [Sun Tzu], a good commander is benevolent and unconcerned with failure" (Episode 502). Following Sun Tzu, the mob is ever watchful of the need for logistics, strategy and tactics to avoid "taking a hit", following the guide that "the best battle is the battle won without being fought".

Sun Tzu's presence in popular culture has ignited another wave of zeal toward *AOW*. According to Sara Leopold, spokeswoman of Oxford University Press in New York, thanks to The Sopranos, the sales of Griffith's translation soared from about 1,000 copies a month to more than 14,000 in the month after episodes of The Sopranos aired and 25,000 more were ordered from the printer to meet readers' demand (Chu 2002; Kaplan 2001). For Westview Press, an academic publisher in Boulder, Colorado, *AOW* became one of the company's three most popular books, with sales about 2,500 a month. It sold out 12,000 copies and demanded an emergency reprint. When the TV series was on, Amazon.com cited *AOW* as the No. 1 bestseller in Bergenfield, N.J.— the center of Soprano territory.

5.4.3 Sun Tzu in other aspects of society

The translated Sun Tzu also infiltrated many other discourses such as law, health, feminism, love, marriage and family.

Sun Tzu is active in legal affairs. Bettler and Spievak (2008), in his journal paper "The Art of War for Litigators", applied Sun Tzu's five-part framework for evaluation before war: the Tao, Heaven, Earth, the Generals and Organization to deliberate the litigation issues. Shepherd and Smith (2013) directly quoted more than nine times from Giles' translation. Sun Tzu's principal ideas, such as knowing the enemy and yourself, attacking the weak and avoiding the strong and moving like water, are used to provide advice for a good trial plan. Other journal papers of this kind include "Trial Warrior: Applying Sun Tzu's *The Art of War* to Trial Advocacy" (Pribetic 2007). A monograph entitled *Sun Tzu and the Art of Litigation: Tipping the Scales of Justice in Your Client's Favour* (Morris-Cotterill 2012) comprises 13 chapters that discuss legal affairs, using Giles' translation of *AOW*. After each sentence of the translation, is the author's exposition on litigation.

In the discourse on health, Sun Tzu is also alluded to. For instance, Fermano (2014) quotes Sun Tzu's indirect approach from Giles' translation to discuss the art of war against cancer. Silver (1995) used the ideas of deception and surprise from Clausewitz and Sun Tzu to reflect the art of health reform in the US. Dunn (2011) not only quoted directly from *AOW*, but also imitated the style of Sun Tzu to propose suggestions for US medicine and physicians in peer review proceedings. During the global outbreak of the COVID-19 pandemic in 2020, Griffith's translation "Every battle is won or lost before it is even fought" is cited at the very beginning of a journal editorial to help analyze the situation and find control measures in the style of Sun Tzu: preparation, positioning, elimination vs mitigation and finally winning the war (Reperant and Osterhaus 2020).

Sun Tzu also appears in feminist works. One book entitled *The Art of War for Women* (Chu 2007) shows women how to use Sun Tzu's philosophy to win in every aspect of life, mainly the principle of victory without fighting, evaluation of strength and weakness, understanding ourselves and other. It consists of 13 chapters, following the structure of that of *AOW*. Sheetz-Runkle (2011) adapted the Sun Tzu strategies to business competition and offered women dozens of straightforward tips on how to be more successful in business. Huang and Rosenberg (2011) also wrote a book *Women and the Art of War: Sun Tzu's Strategies for Winning Without Confrontation*. With a 13-chapter structure similar to that of *AOW*, it covers Sun Tzu's strategic principles point by point to illustrate how women can find their strengths, meet their weaknesses head-on, deal with obstacles and forge their own success in their careers and personal life.

Other books direct Sun Tzu's strategy to the discourse of love. For instance, *The Art of War for Lovers* (Cowan and Parent 1998), based on Sun Tzu, offers people a new perspective of love on how to outsmart opponents and create a long, loving relationship. Gagliardi (2002a) in his book *Sun Tzu's The Art of War Plus The Art of Love*, applied Sun Tzu's tactics to find, win and keep a lifelong love. The layout of the book is rather special. The left-hand pages contain Gagliardi's own translation of *AOW*, including the principle of winning without conflict. The right-hand pages contain a line-for-line adaptation for pursuing lifelong love. *The Art of War for Dating* (Rogell 2010), with lots of direct quotations from Giles' translation and a 13-chapter overall structure, extracts Sun Tzu's key ideas and applies them to dating skills.

Sun Tzu can also be found in family discourse. For instance, Khoo (2002a; 2002b) applied *AOW* in managing marriage and children. Michaelson and Michaelson (2003b) dwelt on how to apply Sun Tzu's strategy to face challenges and accomplish important goals in people's lives. Gagliardi (2002b) also composed a book to offer a guide for raising teenagers with Sun Tzu's text and a line-by-line adaptation.

Sun Tzu's strategies and tactics are also believed to be effective in spiritual life to build up a lasting inner peace (Schnarr 2000), in fiction writing for authors (Bell 2009), and in academic culture and curriculum reform for librarians (Kempcke 2002).

5.5 Summary

The above analysis has shown that Sun Tzu's strategic theory has entered both military and non-military Western discourses. The integration of Sun Tzu in the West has undergone three phases at the individual, institutional and societal levels. First of all, Sun Tzu was examined, compared with Western counterparts and absorbed into Western strategic thinking by individual strategy theorists and masters such as Hart, Handel, Nixon and Boyd. These key figures helped promote the translated Sun Tzu into a higher level of representation in military institutions, especially in the four branches of the US Armed Forces. Next, the translations have been extensively quoted and intensively studied in military doctrines, PME universities and colleges. Sun Tzu also became a guru to guide the wars conducted by US military forces, such as the Gulf War and Iraq War. Thirdly, with the great popularity gained in military sphere, Sun Tzu expanded into many other aspects of social life: business, management, sports, entertainment, law, health and family life. The presence of Sun Tzu in the third phase marks the success of the canonization of *AOW* in Western culture.

The above investigation also displays the fact that translators' choices in core texts and paratexts have played an essential role in the reception of Sun Tzu in the west. To be specific, Giles' literal translation of the water simile led to its resonance to Liddell Hart. The effective translation of *cheng* and *ch'i*, with paraphrasing in the core text and transliteration in the paratexts, inspired Hart's indirect approach theory. Sun Tzu's life story translated from the SL paratexts contributed to the Shock and Awe theory, later tested in Iraq War. The short and basic comparisons between Sun Tzu and Western strategists in Giles' and Griffith's paratexts served as a basis for in-depth and systematic comparisons between the East and West's strategic cultures by later military theorists such as Handel. With concerted efforts, both core texts and paratexts in translations of Sun Tzu have laid a solid foundation upon which Chinese strategic culture can be further studied and applied in Western discourses.

References

Anonymous. 1990. "Sun Tzu's *The Art of War*: 'Commandant's Choice'." *Leatherneck* 73 (7): 39.

Anonymous. 2000. "Sun Tzu Security President and CEO E. Kelly Hansen Appointed to Governor's High Tech Council." *PR Newswire,* December 1, 1. http://search.proquest.com/docview/449237354?accountid=12206.

Anonymous. 2003. "Week in Review." *Milwaukee Journal Sentinel,* January 12, 2D. http://search.proquest.com/docview/261796254?accountid=12206.

Ashley, Chip. 2008. "Sun Tzu: The Art of War and Basketball." *Coach and Athletic Director* 78 (2): 50–52.

Bell, James S. 2009. *The Art of War for Writers: Fiction Writing Strategies, Tactics, and Exercises.* 1st ed. Cincinnati, OH: Writer's Digest.

Berkowitz, Bruce. 2003. "John Boyd: The American Sun Tzu (book review)." *Orbis: A Journal of World Affairs* 47 (2): 370.

Bettler, Robert F., and Elizabeth R. Spievak. 2008. "The Art of War for Litigators." *The Practical Litigator* 19 (1): 51–58.

Borik, Frank C. 1995. *Sub Tzu and the Art of Submarine Warfare*. Research Report. Maxwell Air Force Base: Air War College, Air University.

Boyd, John. 1986. "Patterns of Conflict Briefing." www.d-n-i.net/boyd/pdf/poc.pdf (Accessed December 20, 2015).

Boyd, John. 1987a. "Organic Design for Command and Control." www.d-n-i.net/boyd/pdf/poc.pdf (Accessed December 20, 2015).

Boyd, John. 1987b. "The Strategic Game of? and?" www.d-n-i.net/boyd/pdf/poc.pdf (Accessed December 20, 2015).

Brahm, Laurence J. 2004. *Doing Business in China: The Sun Tzu Way*. 1st ed. Boston: Tuttle.

Brodkin, Jon. 2006. "I Like Books That Get You Thinking – *The Art of War* Is a Great Book." *The Guardian,* April 22. www.theguardian.com/football/2006/apr/22/sport.comment1.

Burke, Steven. 2011. "Hurd Has Done It Again." *CRN,* November 1. http://search.proquest.com/docview/906175090?accountid=12206.

Candela, Harry D. 1998. *"Sun Tzu's The Art of War* in United States Marine Corps Officer Education." MA Thesis. Robertson School of Government, Regent University.

Chapin, Gary P., and T. L. McDonald. 1992. *Sun Tzu's Ancient Art of Golf*. Chicago: Contemporary Books.

Chu, Chin-Ning. 2007. *The Art of War for Women: Sun Tzu's Ancient Strategies and Wisdom for Winning at Work*. New York: Currency Doubleday.

Chu, John. 2002. "Related article: The Art of War and East Asian Negotiating Styles." *Williamette Journal of International Law & Dispute Resolution* 10: 161.

Coram, Robert. 2002. *Boyd: The Fighter Pilot Who Changed the Art of War*. 1st ed. Boston and London: Little Brown.

Cowan, Connell, and Gail Parent. 1998. *The Art of War for Lovers*. New York and London: Pocket.

Critzer, Orlando D. 2012. "21st Century Strategy Needs Sun Tzu." MA Thesis. United States Army War College.

Deva, Mukul. 2005. *S.t.r.i.p.t.e.a.s.e: The Art of Corporate Warfare*. New Delhi, India: Penguin Books India; Viking.

Dunn, John D. 2011. "*The Art of War* Adapted to US Medicine." *Journal of American Physicians and Surgeons* 16 (1): 25–27.

Edelman, Asher. 1987. "Business Forum: Columbia Business School and The Raider's Reward; Train Students in *The Art of War*." *New York Times,* November 22. Retrieved from http://search.proquest.com/docview/426658908?accountid=12206.

Fermano, Olivia. 2014. "The Art of War Against Cancer." *Penn Medicine News*, September 23. http://news.pennmedicine.org/blog/2014/09/the-art-of-war-against-cancer.html.

Fitzsimmons, Scott. 2007. "Evaluating the Masters of Strategy: A Comparative Analysis of Clausewitz, Sun Tzu, Mahan, and Corbett." *Innovations: A Journal of Politics* 7: 27–40.

Franks, Tommy R., and Patrecia S. Hollis. 1991. "1st Cav in Desert Storm—Deception, Firepower and Movement." *Field Artillery* (June): 31–34.

Gagliardi, Gary. 2002a. *Sun Tzu's The Art of War, Plus, The Art of Love: The Ancient Chinese Science of Strategy, Bing-Fa, for Finding, Winning and Keeping Lifelong Love*. 1st ed. *The Art of War Plus*. Seattle, WA: Clearbridge Publishing.

Gagliardi, Gary. 2002b. *The Art of War: Plus, the Art of Parenting Teens*. 1st ed. *The Art of War Plus*. Shoreline, WA: Clearbridge Publishing.

Gagliardi, Gary. 2003. *Sun Tzu's The Art of War: Plus, the Art of Marketing*. 1st ed. *The Art of War Plus*. Shoreline, WA: Clearbridge Publishing.

Garner, Rochelle. 2006. "Oracle and *The Art of War*." *The Ottawa Citizen,* October 20. http://search.proquest.com/docview/241024870?accountid=12206.

Giles, Lionel. 1910. *On The Art of War: The Oldest Military Treatise in the World (Translated from the Chinese with Introduction and Critical Notes by Lionel Giles)*. London: Luzac.

Gillam, Mary M. 1997. "Information Warfare: Combating the Threat in the 21st Century." Research Paper. Air Command and Staff College (US).

Hall, Allan. 2002. "Companies Makes Military Moves on a Corporate Battlefield." *Knight Ridder Tribune Business News,* May 8. http://search.proquest.com/docview/461780856?accountid=12206.

Handel, Michael I. 1991. *Sun Tzu and Clausewitz Compared*. Carlisle Barracks, PA: Strategic Studies Institute, US Army War College.

Handel, Michael I. 1992. *Masters of War: Sun Tzu, Clausewitz, and Jomini*. London: Frank Cass.

Handel, Michael I. 2000. "Corbett, Clausewitz, and Sun Tzu." *Naval War College Review* 53 (4): 106.

Handel, Michael I. 2001. *Masters of War: Classical Strategic Thought*. 3rd ed. London and Portland: Frank Cass.

Harry, Chris. 1997. "In Sun's Shadow: Gator's Spurrier Learns from A Chinese Warrior." Sep 16. http://search.proquest.com/docview/388433597?accountid=12206 (November 16, 2020).

Hart, Liddell. 1954. *Strategy: The Indirect Approach*. New York: Frederick A Praecer.

Hart, Liddell. 1963. "Foreword." In *Sun Tzu: The Art of War*, trans. Samuel B. Griffith, vi–vii. New York: Oxford University Press.

Hart, Liddell. 1929. *The Decisive Wars of History: A Study in Strategy*. London: G. Bell & Sons ltd.

Hawkins, David E., and Shan Rajagopal. 2005. *Sun Tzu and the Project Battleground*. New York: Palgrave Macmillan.

Headquarters US Marine Corps. 1989. *FMFM 1 Warfighting*. Washington D.C.

Headquarters US Marine Corps. 1990. *FMFM 1–1 Campaigning*. Washington D.C.

Headquarters, Department of the Army (HDA). 1982. *Field Manual 100–5, Operations*. Washington D.C.

Headquarters, US Air Force Doctrine Center. 1999. *Air Force Doctrine Document 2–5.3*. Maxwell Air Force Base: Air Force Doctrine Center.

Holmes, James. 2003. "Sun Tzu's Ideas Had Major Influence in Operation Iraqi Freedom Strategy." *Navy Times* (June 2): 54.

Huang, Catherine, and Arthur D. Rosenberg. 2011. *Women and the Art of War: Sun Tzu's Strategies for Winning Without Confrontation*. 1st ed. Tokyo and Rutland, VT: Tuttle Publishing.

Hwang, Ching-Chane, and L. H. M. Ling. 2009. "The 'Kitsch' of War: Misappropriating Sun Tzu for an American Imperial Hypermasculinity." In *Gender and Global Politics in the Asia-Pacific*. 1st ed., eds. Bina D'Costa and Katrina Lee-Koo, 59–76. New York: Palgrave Macmillan.

Hyde, Marina. 2003. "Comment & Analysis: This Week." *The Guardian*, March 29(1): 22.

Ipton, Martin. 2006. "How Big Phil Plans to Win the Football War." *The Daily Mirror,* June 28. http://search.proquest.com/docview/339832659?accountid=12206.

Johnson, Eric N. 1992. *Sun Tzu in the Age of Technology.* Research Report. Maxwell Air Force Base: Air War College, Air University.

Johnston, Alastair I. 1999. "Sun Zi Studies in the United States." (unpublished report) Cambridge, MA: Harvard University.

Kaplan, Fred. 2001. "A Soprano Speaks And `War' Erupts Sun-Tzu's Classic Gets Boost from HBO Series." *Boston Globe,* May 5. http://Search.Proquest.Com/Docview/405409879?Accountid=12206v.

Kempcke, K. 2002. "The Art of War for Librarians: Academic Culture, Curriculum Reform, and Wisdom from Sun Tzu." *Portal* 2 (4): 529–551.

Khoo, Kheng-Hor. 2002a. *Applying Sun Tzu's Art of War in Managing Your Marriage.* Selangor Darul Ehsan, Malaysia: Pelanduk Publications.

Khoo, Kheng-Hor. 2002b. *Applying Sun Tzu's Art of War in Selling. Sun Tzu's Art of War.* Selangor Darul Ehsan, Malaysia: Pelanduk Publications.

Krause, Donald G. 1995. *The Art of War for Executives.* London: Nicholas Brealey.

Lawlor, Julia. 1991. "Ancient Warrior in Persian Gulf." *USA Today,* February 22. http://search.proquest.com/docview/306437478?accountid=12206.

Lee, Bradford A. 2010. "Teaching Strategy: A Scenic View from Newport." In *Teaching Strategy: Challenge and Response*, ed. Gabriel Marcella, 105–148. Carlisle, PA: Strategic Studies Institute, US Army War College.

Leonhard, Robert R. 2003. "Sun Tzu's Bad Advice: Urban Warfare in the Information Age." *Army* 53 (4): 38–42, 44.

Lord, Carnes. 2000. "A Note on Sun Tzu." *Comparative Strategy* 19 (4): 301–307.

Macintyre, Ben. 2003. "They Fought by the Book, and It Was Sun Tzu Who Won It." *The Times*, April 12. http://search.proquest.com/docview/318859092?accountid=12206 (Accessed February 18, 2020).

Marshall, Andrew S. 1999. "Is the Sun Tsu's Art of War Relevant to the Modern Air Force?". *RUSI Journal* 144 (5): 57–60.

McCann, Dennis P. 2012. "On Reading Sun-Tzu: The Promise and Perils of Appropriating a Chinese Classic in International Business Ethics." *Journal of International Business Ethics* 5(2): 27–37.

McElhatton, Emmet J. 2014. "Professional Reading and the Education of Military Leaders." PhD Thesis. Victoria University of Wellington.

McNeilly, Mark. 1996. *Sun Tzu and the Art of Business: Six Strategic Principles for Managers.* New York: Oxford University Press.

McNeilly, Mark, and Mark R. McNeilly. 2001. *Sun Tzu and the Art of Modern Warfare.* New York: Oxford University Press.

Mendez, Anthony W. 1995. *The Gettysburg Campaign: "Know Your Enemy, Know Yourself."* Research Report. Maxwell Air Force Base: Air War College, Air University.

Michaelson, Gerald A. 1998. *Sun Tzu: The Art of War for Managers: 50 Rules for Strategic Thinking.* Avon: Adams Media.

Michaelson, Gerald A., and Steven Michaelson. 2003a. *Sun Tzu: Strategies for Selling: How to Use The Art of War to Build Lifelong Customer Relationships (translations from the Chinese by Pan Jiabin and Liu Ruxian).* New York: McGraw-Hill.

Michaelson, Gerald A., and Steven Michaelson. 2003b. *Sun Tzu for Success: How to Use The Art of War to Master Challenges and Accomplish the Important Goals in Your Life.* Avon, MA.: Adams Media Corp.

Michaelson, Gerald A., and Steven Michaelson. 2004. *Sun Tzu Strategies for Marketing: 12 Essential Principles for Winning the War for Customers.* New York and London: McGraw-Hill.

Millett, Allan R. 1991. *Semper Fidelis: The History of the United States Marine Corps. The Macmillan Wars of the United States.* New York: Free Press; Toronto: Maxwell Macmillan Canada; New York: Maxwell Macmillan International.

Morgan, Matthew J. 2005. "The Art of War in Operation Iraqi Freedom." *Defense & Security Analysis* 21 (1): 97–104.

Morris-Cotterill, Nigel. 2012. *Sun Tzu and the Art of Litigation.* Scotts Valley, CA: Nigel Morris-Cotterill.

NATO Standardization Agency. 2013. *AAP-06 NATO Glossary of Terms and Definitions.* Brussels: NATO Standardization Office.

Neilson, Robert E., ed. 1997. *Sun Tzu and Information Warfare: a Collection of Winning Papers from the Sun Tzu Art of War in Information Warfare Competition.* Washington, D.C.: National Defense University Press.

Nisse, Jason. 1999. "Don't Give Your Rivals a Fighting Chance." *The Times,* March 27, 30. http://search.proquest.com/docview/318080226?accountid=12206.

Nixon, Richard. 1980. *The Real War.* New York: Warner Communication Company.

O'Dowd, Edward, and Arthur Waldron. 1991. "Sun Tzu for Strategists." *Comparative Strategy* 10 (1): 25–36.

Osinga, Frans P. B. 2007. *Science, Strategy and War: The Strategic Theory of John Boyd.* Vol. 18 of *Strategy and History.* London: Routledge.

Paquette, Laure. 1991. "Strategy and Time in Clausewitz's *On War* and in Sun Tzu's *The Art of War.*" *Comparative Strategy* 10 (1): 37–51.

Phua, Chao Rong Charles. 2007. "From the Gulf War to Global War on Terror -a Distorted Sun Tzu in US Strategic Thinking?" *RUSI Journal: Royal United Services Institute for Defence Studies* 152 (6): 46.

Pribetic, Antonin I. 2007. "Trial Warrior: Applying Sun Tzu's *The Art of War* to Trial Advocacy." *The Alberta Law Review* 45: 1017–1035.

Rasmussen, Mikkel V. 2001. *The Acme of Skill: Clausewitz, Sun Tzu and the Revolutions in Military Affairs.* Copenhagen: Dansk Udenrigspolitisk Institut (DUPI).

Reperant, Leslie A., and Albert D. M. E. Osterhaus. 2020. "COVID-19: Losing Battles or Winning the War?" *One Health Outlook* 2 (1): 1–3.

Rhem, Kathleen T. 2002. "Major Publishers Join to Send Books to Troops." US Department of Defense. November 7. http://archive.defense.gov/news/newsarticle. aspx?id=42527 (December 7, 2019).

Rice, Sean P. 2006. "Sun Tzu: Ancient Theories for a Strategy Against Islamic Extremism." Research paper. US Army War College.

Richards, Chester W. 2003. *A Swift, Elusive Sword: What if Sun Tzu and John Boyd did a National Defense Review?* 2nd ed. Washington, D.C.: Center for Defense Information.

Richter, Paul. 1991. "Ancient Doctrine Guiding Futuristic Warfare in Gulf: Strategy: Sun Tzu's Tiny Book, Written More than 2,500 Years Ago, Is Influencing U.S. and Iraqi Tactics." February 18. www.latimes.com/archives/la-xpm-1991-02-18-mn-1191-story.html (Accessed June 21, 2020).

Roberson, Darryl L. 2002. "Fight Against Terrorism Sun Tzu Revisited." Course paper for Course 5602. National War College.

Rogell, Eric. 2010. *The Art of War for Dating: Master Sun Tzu's Tactics to Win over Women*. Avon MA.: Adams Media.

Romm, Joseph J. 1991. "The Gospel According to Sun Tzu." *Forbes* 148 (13): 154–162.

Saunders, Peirce C. 2009. "Sun Zi's *Art of War* and US Joint Professional Military Education." *INSS Proceedings:* 1–2. www.ciaonet.org/catalog/18014.

Schnarr, Grant R. 2000. *The Art of Spiritual Warfare: A Guide to Lasting Inner Peace Based on Sun Tzu's The Art of War (Foreword by Robert L Moore)*. 1st ed. Wheaton, IL., London: Theosophical Publishing House.

Sheetz-Runkle, Becky. 2011. *Sun Tzu for Women: The Art of War for Winning in Business*. Avon, MA: Adams Media.

Sheetz-Runkle, Becky. 2014. *The Art of War for Small Business: Defeat the Competition and Dominate the Market with the Masterful Strategies of Sun Tzu*. New York: AMACOM.

Shepherd, William N., and Thomas D. Smith. 2013. "Sun Tzu and the Art of Trial." *Litigation* 39 (1): 24–27.

Silver, George A. 1995. "Topics for Our Times: Clausewitz vs Sun Tzu-The Art of Health Reform." *American Journal of Public Health* 85 (3): 307–308.

Sloan, Elinor C. 2012. *Modern Military Strategy: An Introduction*. London and New York: Routledge.

Stein, George J. 2007. "CE 6558 Sun Tzu, The 'Seven Military Classics' and Unconventional Strategic Thought." Course Syllabus. Air War College.

The Joint Doctrine and Concept Center UK. 2001. *Joint Warfare Publication 0–01 British Defense Doctrine*. 2nd ed. Swindon.

Thibault, George E. 1984. *The Art and Practice of Military Strategy*. Research Report. Washington D.C.: National Defense University.

TUSI. 1963. "Sun Tzu: *The Art of War*." *Royal United Service Institute Journal* 108: 185–186.

Ullman, Harlan. 2003. "'Shock and Awe' Misunderstood." *USA TODAY,* Apr 08.

Ullman, Harlan K. 2010. "Shock and Awe a Decade and a Half Later, Still Relevant, Still Misunderstood." *Prism* 2 (1): 79–86.

Ullman, Harlan K., and James P. Wade. 1996. *Shock and Awe: Achieving Rapid Dominance*. Washington D.C.: National Defense University.

US Joint Chiefs of Staff. 1998. *Joint Pub 3–13 Joint Doctrine for Information Operations*. Washington, DC.

Wade, Don. 2006. *Golf and the Art of War: How the Timeless Strategies of Sun Tzu Can Transform Your Game*. New York: Thunder's Mouth Press.

Wheeler, Gavin. 2014. "The Great War's Leadership Lessons." *City A.M.,* October 27, 25. http://search.proquest.com/docview/1616412349?accountid=12206.

Wilcoxon, Gregory L. 2010. "Sun Tzu: Theorist for the Twenty-First Century." Strategy Research Project. US Army War College.

Winter, Henry. 2006. "Scolari Schooled in *The Art of War*: Ancient Chinese Text Key to Portugal Coach's Strategy for England Tie." *The Daily Telegraph,* June 28. http://search.proquest.com/docview/321333450?accountid=12206.

Wojciechowski, Gene. 1992. "Knight Takes His Hoosiers Beyond Mere Will to Win College Basketball: Coach Stresses the Will to Prepare to Win in Discussing His Philosophy, the NCAA and *The Art of War*." *Los Angeles Times,* January 14. http://search.proquest.com/docview/281535588?accountid=12206.

Yeh, Raymond T.-Y., and Stephanie H. Yeh. 2004. *The Art of Business: In the Footsteps of Giants*. Olathe CO: Zero Time Publishing.

Yuen, Derek M. C. 2014. *Deciphering Sun Tzu: How to Read The Art of War*. New York: Oxford University Press.

Zeman, Ned. 2011. "Michael Ovitz, Take Two." June 21. www.vanityfair.com/news/2001/04/ovitz-200104 (June 21, 2020).

Zweibelson, Ben. 2010. "The US in Afghanistan Follow Sun Tzu Rather than Clausewitz to Victory." *Small Wars Journal* (December): 1–6.

6 Social-cultural context of the translation and reception of Sun Tzu

Our textual analysis of the two translations of *The Art of War* has shown that with concerted efforts with the core texts and paratexts, Sun Tzu's strategic thinking can be fluently translated and the ancient Chinese strategic culture effectively reconstructed. The investigation of the reception of Sun Tzu's work aided by the self-built corpus of 410 military texts reveals that Chinese strategic culture has been well-received and has led to Western (especially US) strategic innovation. A survey of the self-built Corpus 3 of more than 80 texts shows that *AOW* has transcended the military boundary to reach and influence other fields of society. The re-canonization of Sun Tzu in the Western discourse is evident. We might continue to ask: Why did Giles and Griffith make such strenuous efforts to reconstruct this strategic culture? And, more importantly, why was the translated Sun Tzu so greatly welcomed by Western culture? In this section, we will discuss these questions from the following aspects: the immediate situational constraints (including the translator's identity and the target readership), the military demand for strategic innovation, and power structure and ideology in the target culture.

6.1 Immediate situational constraints

It is commonly believed that translations are influenced by the immediate situational contexts in which they are done. In this section, we will probe into the translator's social identity and the target readership, the two most important factors that are directly related to the translator's treatment of the core text and paratexts.

6.1.1 Translator's social identity

In this part, we will look at the social identities of the two translators, Giles and Griffith, so as to explain why the core texts and paratexts are treated as they are in the translations.

Giles was a leading sinologist in his time, representing "an era of sinology when a scrupulous respect for and familiarity with ancient texts was combined with a broad reading in several European languages, engagement with major

DOI: 10.4324/9781003025726-6

intellectual issues and trends of the day, and a fluent English prose style" (Minford 2008, xv). His education in Classics at Oxford University enriched his understanding of the importance of annotation for classic works and his knowledge of Western cultural heritage. Giles inherited a great enthusiasm for Chinese culture from his father, Herbert Giles (1845–1935), a British diplomat, sinologist, and professor of Chinese language. His long period of work from 1900 to 1940 as Keeper of the Department of Oriental Manuscripts and Printed Books in the British Museum helped him to become well acquainted with ancient Chinese culture. He boasted a long list of translations of Chinese classics, including *The Sayings of Lao Tzu* (1904), *The Sayings of Confucius* (1907), and *Taoist Teachings from the Book of Lieh Tzu* (1912). He held Chinese culture in such a high esteem that he once confessed himself as a Taoist (Minford 2008, xvi). Therefore, it was natural for him to express through his notes his respect toward ancient Chinese military culture.

Giles' identity as a sinologist determined the way he dealt with the core text and paratexts. Firstly, his main purpose was to promote cultural exchange between East and West. The extensive citations from books of Western military culture in Giles' notes show that he was well-versed in both Eastern and Western strategic culture before he started the translation of Sun Tzu. Giles identified mainly the similarities between the two strategic traditions. Although Giles' expertise in strategy was not so deep as to enable him to discern the major differences between Eastern and Western strategic culture, Giles displayed his great admiration of Sun Tzu and successfully aroused resonance among his readers, including Hart. His mission as a sinologist to spread Chinese strategic culture was substantially successful. Secondly, as an admirer of Taoism, he stressed one of his main arguments, that *AOW* is a peace-oriented military treatise, in his paratexts. However, such efforts in bringing out the peace-orientation seemed to be ignored by most of the Western military theorists who studied his translation.

Rather different from Giles, Griffith (1906–1983) was a brigadier general in the US Marine Corps. After graduating from the US Naval Academy, Griffith entered the Marines in 1929 and served as an officer and commander until his retirement in 1956. For his excellence, he received the awards of the Navy Cross in 1942 and the Army Distinguished Service Cross in 1943. Griffith also lectured widely, including at such establishments as the Armed Forces Staff College, the US Military Academy, the Foreign Policy Association and the Marine Corps Schools.

Griffith had a great interest in and a familiarity with Chinese language and military affairs. His first stay in China lasted three years, from June 1935 to July 1938. For the first two years, he was a diligent Chinese language student at the American Embassy in Peking, spending six hours a day in instruction. During his second stay in China, from January 1946 to May 1947, Griffith commanded the 3rd Marine Regiment and later the US Marine Forces in Qingdao after participating in the occupation of North China after World War II.

Griffith was a prolific writer and translator of military history. Specifically, he was an expert in Chinese military strategy, with many papers published in the *Marine Corps Gazette*, the *Saturday Evening Post*, *Naval Institute Proceedings* and the *New Yorker*. Some of these papers are: "North China, 1937" (Griffith 1938), "The Glorious Military Thought of Comrade Mao Tse-Tung" (Griffith 1964), "Communist China's Capacity to Make War" (Griffith 1965) and "Chinese Capabilities and Strategies for the 1970's" (Griffith 1969). His books include *The Battle of Guadalcanal* (1963), *Peking and People's Wars* (1966), *The Chinese People's Liberation Army* (1967), *War for American Independence* (1976) and *History of the Second World War* (1974).

Griffith's translation of Mao Tse-tung's *On Guerrilla War* was published originally under the title "Guerrilla Warfare in China" in the *Marine Corps Gazette* as a journal article in June 1941, and was reprinted by Praeger Publishers as a book in 1961. Griffith notes that:

> The influence of the ancient military philosopher Sun Tzu on Mao's military thought will be apparent to those who have read "The Book of War". Sun Tzu wrote that speed, surprise and deception were the primary essentials of the attack and his succinct advice "Sheng Tung Chi His" (Distraction in the East, Strike in the West) is no less valid today than it was when he wrote it twenty-four hundred years ago. The tactics of Sun Tzu are in large measure the tactics of China's guerrillas today.
>
> (Griffith 1941, 19)

In 1943, another article by Griffith appeared in the same journal, "That Man Suntzu". With many direct quotations from Calthrop's translation of Sun Tzu in 1908, Griffith tried to explain how the Japanese army employed the principles of Sun Tzu to launch an attack against America at Pearl Harbor in 1941 (Griffith 1943).

Upon his retirement in 1956, Griffith enrolled in New College, Oxford University, as a PhD student under the tutorial guidance of Dr. Wu Shih-ch'ang. He finished his doctoral thesis, the translation of Sun Tzu, in 1960 and was awarded a PhD in Chinese military history in 1961. Based on a significant revision of his doctoral thesis, Griffith's translation of *AOW* was published by Oxford University Press in 1963. Griffith's journal paper "Some Chinese Thoughts on War" began to cite his own translation of Sun Tzu (Griffith 1961). He outlined ten principles from Sun Tzu, including morale, deception, surprise, mobility, timing, disruption and flexibility, are portrayed as a framework of four pillars of the contemporary Chinese army: intelligence, estimates, planning and prudence.

Griffith's identity as a US marine general and military scholar had a great influence on his translation of *AOW* and its reception. Firstly, his main purpose was to foster an understanding of Chinese military thinking among the US armed forces and politicians during the Cold War period. Secondly, as a researcher of Mao Tse-tung's guerrilla warfare, he aimed to discern the main connection between Sun Tzu and Mao Tse-tung. This intention is revealed in his paratext about Mao Tse-tung. Thirdly, what mattered most for Griffith were the strategies to win the war but not to promote pacifism. Therefore, in his paratexts, the peace-orientation of Sun Tzu is somewhat neglected. Clearly, Griffith's prestigious status as a military expert, strengthened by his extensive publication in military spheres, partly guaranteed his translation a greater popularity and wider reception than Giles'.

6.1.2   Target readership

Another important factor impacting the translations of *AOW* is the readership for whom the translators were working. For instance, the use of annotation, according to Delisle (1999:196), is "a sign of intellectual integrity and conscientious professionalism on the part of translators who are concerned about the needs of the target audience and wish to enhance the readability of the target text".

It is almost certain that Giles' earliest target readers included his two brothers. One of them was Lancelot Giles (1878–1934), appointed in 1900 as a young student interpreter in Peking, where he was decorated for military gallantry in the defense of the Western army legation involved in fighting the Chinese during the Boxer Uprising (see Minford 2008, xxii). The other was Valentine Giles (1877–1945), who became a captain in the Royal Engineers in 1908. On the title page of his translation of *AOW*, Giles stated: "To my young brother, Captain Valentine Giles, RC., in the hope that a work 2,400 years old may yet contain lessons worth consideration to the soldier of today, this translation is affectionately dedicated". His two brothers should have benefited from the strategic thinking of Sun Tzu. This dedication also suggests that Giles expected many other British officers and soldiers to profit from reading his translation. Naturally, in his translation of the core text, its register as a military thesis was kept as were the majority of the supplemented paratexts involving military issues.

Furthermore, it is also conceivable that Giles, as a sinologist, wanted to welcome into the army of his readers those who were interested in but seldom exposed to Chinese culture. In order to help them better comprehend the ancient, alien and terse Chinese strategic theory, Giles turned to his notes of Western culture. At the same time, he wanted to fascinate his readers with a treasured masterpiece and consequently added a large number of commendatory notes to foster a due respect for Sun Tzu.

The target readership of Griffith's translation is quite different from that of Giles'. According to Johnston (1999), it was discovered that Griffith once asked the publisher of his translation to distribute copies to top leaders at the Department of Defense, Department of State and the White House, as well as to key journalists and opinion-makers. Griffith saw his translation as a tool to influence senior US military and political leaders about how to deal with revolutionary warfare in the Third World. His purpose was to warn US strategic decision makers about the approaches that China, North Vietnam and other revolutionary states were taking to threaten US interests. These strategic decision makers would be more interested in contemporary Mao Tse-tung's military thought, a key inspiration for revolutionary guerilla war movements in the Third World, than in Sun Tzu's. However, they seldom know that Sun Tzu the source of inspiration for Mao's strategy. This partially explains why Griffith included paratexts such as "Sun Tzu and Mao Tse-tung" and "Sun Tzu's Influence on Japanese Military Thought" in his translation. Griffith's target readers would also include contemporary US soldiers and military officers. This is the reason why he contextualized Sun Tzu's thinking in modern settings and used phrases such as "national strategy" and "military strategy" in the paratexts.

6.2 The Western military need for strategic innovation

As our analysis about the reception of the translated Sun Tzu in Western military discourse shows, *AOW* had a drastically increasing popularity among Western strategists after the two world wars. This is largely due to the constant demands for strategic innovation in Western military sphere.

6.2.1 The aftermath of two world wars

The two world wars were conducted within the military theoretic framework structured by Clausewitz's *On War*, with a new focus on technology which magnifies merely the power of killing: "[p]ut in strategic terms, the Western practice that the destruction of the enemy's armed forces and battle are conceived as the only sound aim in war and the only goal of strategy" (Yuen 2014, 131).

World War I, one of the deadliest conflicts in history, ended in 1918 with a death toll of more than 9 million combatants and 7 million civilians, a casualty rate amplified by the belligerents' technological and industrial development and tactical stalemate. According to Handel (2001, 15), during the First World War the Western mind was occupied by the military theories put forward by Clausewitz, who regarded deception and surprise as largely impracticable, but stressed attrition warfare featuring trench-to-trench fighting, stalemate and immobility which resulted in senseless carnage. Eventually, the aftermath First World War presented a necessity for new strategic solutions. Yuen (2014, 131) held that "after the bloodbaths in the First World War, it became apparent that the West was urgently in need of an alternative strategic model that offered better prospects of attaining more certain, and less costly, victories".

Liddell Hart had a clear view of the shortcomings of Western strategy when he was fighting in 1916 for Great Britain, which experienced a loss of 60,000 soldiers in a single day. He strongly opposed the reliance on Clausewitz's strategic guidance. When he encountered Giles' translation of *AOW*, he found something new in the Chinese strategic thoughts which propose a fundamentally different concept of efficacy. Therefore, Sun Tzu's emphasis on attacking weakness and winning without fighting seemed to offer an alternative to the bloody frontal attacks of World War I and contributed to Hart's formulation of the concept of the "indirect approach".

However, Western strategists in the Second World War almost ignored Sun Tzu's maxim of winning without fighting and Hart's indirect approach. World War II, still adhering to Clausewitz's teachings, became the deadliest military conflict in history, with over 60 million people killed. Reflecting the carnages of the two world wars, Hart remarked that:

> Civilization might have been spared much of the damage suffered in the world wars of this century if the influence of Clausewitz's monumental tomes *On War*, which moulded European military thought in the era preceding the First World War, had been blended with and balanced by a knowledge of Sun Tzu's exposition on "*The Art of War*".
>
> (Hart 1963, vi)

The two world wars have warned strategists that mankind's extinction is near if no strategic innovation is made. It is under this circumstance that Sun Tzu was introduced to an increasing number of strategists in a wider scope of research.

6.2.2 The legacy of the Cold War

The Cold War (1945–1989) once again increased the doubts about the sole dependence on military might and generated a call for a revolution in Western grand strategy. The Cold War was the sustained state of political and military tension between powers in the Western Bloc (the United States with NATO and others) and powers in the Eastern Bloc (the Soviet Union and its allies in the Warsaw Pact). It was "cold" because there was no large-scale fighting directly between the two sides despite several regional proxy wars.

In the period, the US failures in the Korean War (1950–1953) and the Vietnam War (1956–1975) further highlighted the need for reflection on the Western traditional strategy. In the Korean War, the military power, superior technology and economic might of the United Nations forces led by US did not bring about a decisive victory. Similarly, in the Vietnam War, the US failed to achieve ultimate victory even though the US military won almost all the major engagements on the battlefield. The failures in two regional conflicts spurred the US military establishment to rethink its basic military doctrine and reflect on the differences between its strategy and that of its adversaries. The US leaders looked for answers in the classical works on strategy and war.

Stuart and Tow (1981) found that, in the Korean War, the Chinese had followed the strategy of military deception which can be traced back to Sun Tzu. First, about 300,000 Chinese People's Volunteers (CPV) troops were secretly moved and installed in North Korea before they engaged in hostilities. Then, when the United Nations (UN) forces initiated the "Home by Christmas" offensive, and spread out to push north rapidly, the CPV troops effected a significant surprise counteroffensive by "luring the enemy to penetrate deep", dividing and enveloping the UN forces.

Officers such as Michael Wyly, former vice president of the US Marine University and a former combat officer who served two tours in Vietnam, reflected that the North Vietnamese used a lot of Sun Tzu in their tactics to overcome superior American weaponry (Richter 1991). Ho Chi Minh, Vo Nguyen Giap and many other Vietnamese leaders were faithful students of both Sun Tzu and Mao Tse-tung. In the Vietnam War, Sun Tzu's thinking made sense at a strategic level, as well as in battlefield tactics. A commonly used North Vietnamese tactic was to begin an attack with mortar shelling, then infiltrate a small number of troops behind US lines to sow confusion and set up a strike from the rear. "This followed precisely Sun Tzu's advice about seizing the enemy's attention with the application of a 'direct' force, or *cheng*, then knocking him off balance with an 'extraordinary' force, or *ch'i*" (Richter 1991). The North Vietnamese followed Sun Tzu's lesson of understanding the psychology of both parties in war and nurturing good relations with civilians. Facing massive US bombing, they avoided the strength, and struck the weakness, and allowed the US to wear itself out. Some believed that the success of the North Vietnamese came from Mao's guerrilla strategy, which is also rooted in Sun Tzu's theory. Vietnam had used the lessons of *AOW* with great systematic and devastating effect (Marshall 1999, 57).

It is clear that the application of Sun Tzu's principles by the North Vietnamese was superior to the strategy being used by the US for the situation in Vietnam. Vetter (1996, 17) recommended that all military planners be trained in the differences between the Western strategy of war (as expressed in Clausewitz's *On War*) and the Eastern military teachings found in *AOW*. This new direction of inquiry eventually inspired a significant change at the highest levels of the American military and political leadership.

The nuclear arms race, another significant issue during the Cold War, also demanded the evolution of grand strategic thinking. The two superpowers, the US and the Soviet Union, did not engage directly in full-scale armed conflicts but were both armed heavily in preparation for an all-out nuclear World War III. They followed the doctrine of mutually assured destruction (or MAD): each side had an appalling nuclear arsenal that deterred an attack by the other side, on the basis that such an attack would lead to the total destruction of the attacker. The development of the two sides' nuclear warheads and deployment of conventional military forces almost drew the world to the brink of extinction.

Such a great risk alerted people to the need for a new approach to war theory. In this context, the limitations of Western strategic thinking, magnified by technological innovations, once again become evident. Therefore, more and more Western strategists turned their eyes to the Eastern military theory represented by *AOW*. This is what Nixon wrote in his book *Real War:* to identify the faults of MAD and propose an indirect approach to check Soviet Union weakness with US strength.

6.2.3 *The challenges of unconventional and new wars*

Another cause for the popularity of translated Sun Tzu in Western military discourse since the end of Cold War, especially after the 9/11 terrorist attacks, is the challenge posed by unconventional and asymmetric warfare, particularly the global war on terrorism (GWOT) and cyber war.

With the collapse of the Soviet Union in the wake of the Cold War, the US became the world's only superpower. However, the terrorist attacks on September 11, 2001, revealed a newly emerged pressing threat against US security: asymmetric warfare in the form of terrorism on a global scale. Asymmetric warfare is a new threat to humanity (Homayoun 2004). Unlike the traditional warfare "conducted by the legitimate military forces of nation-states, wherein the objective is either terrain- or enemy-focused", asymmetric warfare is "population-centric nontraditional warfare" waged between a militarily superior power and an inferior power or non-nation-state (Buffaloe 2006, 1,17). It comes out in many aspects and approaches: terrorism, insurgency, an information war, bioweapons attacks and cyber-attacks on the Internet. Asymmetric warfare can be non-militarised and amorphous, fought between a formal military and an informal, less equipped, undermanned but resilient opponent.

With Sun Tzu's tactics used, and their relevance and validity testified to, by US forces in traditional and conventional warfare in the Gulf War and the first phase of the Iraq War, Western strategy analysts were encouraged to further explore Sun Tzu's relevance in asymmetric and unconventional warfare. Williams (2003) stated that when coping with the challenges of asymmetric approaches to warfare, it is fortunate to have Clausewitz and Sun Tzu's wisdom to help guide that change and provide direction to the preparation for asymmetric war and the transformation of the "American Way of War". Homayoun (2004) remarked that the theories of von Clausewitz are mostly applicable to conventional war fighting, whereas "the principle theories of Sun Tzu are equally applicable to conventional as well as asymmetric warfare".

Coker (2003) suggested that in the war on terrorism, *AOW* may well be one of the most important texts of all. US military can benefit from Sun Tzu's strategic ideas of using all resources—economic, social and political—to attain a military purpose, and of preserving the enemy as a means essential to success. Comparing Sun Tzu's tactical advice with al Qaeda's battlefield tactics and training regimens, Bartley (2005, 246) found some clear parallels between the "art of terrorism" and the principles of Sun Tzu. He believed that "al Qaeda is a model of Sun Tzu's principles on indirect warfare", and suggested that by studying *AOW*, Western leaders could not only gain a sharp insight into the shadowy world of Islamist terrorism, but also acquire important strategies necessary for America to win the global war on terrorism. According to Rice (2006), the war on terrorism requires the innovative development of strategies beyond conventional thoughts because of a new kind of enemy, usually unseen and driven ideologically. Sun Tzu's indirect approach to coerce or defeat ones' enemy can be used as an analytical tool to examine the current threat of Islamic extremism in aspects of its foundation, methods, goals and environment, and to assess the current national strategy for fighting terrorism.

Recently, the situation in the cyber war also calls for Sun Tzu's involvement. Cyber attacks can lead to economic damage, physical destruction and even the loss of human life. They are made in a new warfare domain consisting of information, communication networks, and computers and servers. As cyberspace is intangible and the attackers are hidden and anonymous, it is difficult to claim victory or calculate damage. Therefore, fresh perspectives are needed for military leaders today to understand and manage the new threat to national security. Once again, Sun Tzu's *Art of War* came into their vision.

Sun Tzu's warfare strategies are quite applicable for information security. According to Miller (2001), five areas of Sun Tzu's teachings—namely, leadership, soldiers, rules, weapons, and war plans—should be heeded when creating a new security program for an institution. Geers (2011) also believed that *AOW* provides a useful framework for the management of cyber war. Judging from the typology of cyber conflict, classic strategists such as Sun Tzu can provide options for policy makers (Greathouse 2013). Other experts have argued that the US has failed to apply Sun Tzu's lessons properly in the cyber arena, and mistakes must be corrected (e.g. Lunas 2011).

Most recently, Sun Tzu's strategic principles have used to analyze space war and hybrid warfare. Hybrid warfare will occur when future adversaries mix and match forms and modes of warfare to offset conventional military battlefield power.

Western military theorists are worried about two types of threats that may test their technological supremacy on the battlefield. The first threat is that of future adversaries merging types of warfare—such as ethnic conflict, terrorism and regular warfighting—to create overwhelming complexity. The second threat is from "non-trinitarian" adversaries who seemingly cannot be conquered in "Clausewitzian" terms by means of a conventional decisive military operation or battle; such adversaries include a wide range of overt and covert military, paramilitary and irregular forces, terrorists and civilians. However, hybrid war "exhibits several of Sun Tzu's concepts", thanks to, for instance, Sun Tzu's use of both regular and irregular actions to defeat the enemy and employment of both asymmetric and unconventional means to weaken the enemy (Bingöl 2017, 113). As a result, many strategic thinkers are resorting to Sun Tzu's thought for strategic analysis and solutions (Atkinson 2018; Manko and Mikhieiev 2018; Monaghan 2019).

According to Szymanski (2020), Sun Tzu's ancient principles are relevant for today's space warfare. Since space warfare thinking is still in its infancy, simply applying Sun Tzu's maxims into a space warfare strategy could prove decisive in a future space battle. Therefore, more than a dozen *AOW*-guided principles for space warfare are listed by Szymanski (2020, 81–82): "Study an adversary's space doctrine, strategies, tactics, organizations, and leadership personalities to discover his strengths and weaknesses so you may better catch him off guard during space systems surprise attacks"; "Continually harass the fixed space systems defenses of your adversaries so they are constantly off-balance, more hurried, and less timely in fulfilling their mission objectives"; and "You may sacrifice some space assets to make your adversaries believe in your carefully falsified military objectives" and so on.

6.3 Power and ideology

In addition to these situational and institutional context factors, power and ideology are two hidden constraints on the translation and reception of *AOW* in the social-cultural context, according to the main arguments of the CDA.

6.3.1 Power inequality and competition

The term "strategic culture" is now an indispensable part of the vocabulary in national security and international relations. However, it should be noted that the strategic culture concept is deeply rooted in the power inequality and the superpower rivalry between the East and the West. In this case, it is in the context of this power competition that Sun Tzu's *AOW* has been received as a canonical work of Chinese strategic culture and has been incorporated into its Western counterpart.

With the US involvement in Vietnam and the US–Russian nuclear confrontation of the Cold War, it became increasingly evident that a coherent concept was needed to understand why countries contemplate violence and wage wars in different ways (Zaman 2009, 71). RAND analyst Jack Snyder launched the strategic culture movement in 1977. He introduced the term "strategic culture" in a research report to analyze the Soviet limited nuclear warfare doctrine. He defined it as: "the sum total of ideals, conditional emotional responses, and patterns of habitual behaviour that members of the national strategic community have acquired through instruction or imitation and share with each other with regard to [nuclear] strategy" (Snyder 1977, 5). Later, Snyder (1990, 3) explained why he promoted the idea of strategic culture: it grew from his realization that the Soviets approached the key questions of strategy in the nuclear era from a viewpoint that was distinct from the United States' doctrine of fighting a limited nuclear war.

Many experts began to look at the different ways in which China, the US, the Soviet Union and Britain fought wars through the lens of strategic culture. Johnston (1995, 5) summarized three generations of the research on strategic culture. His summary revealed the fact that strategic culture is studied as a basis to reflect on the competition for military power. This is also true of the translation and reception of *AOW*. The popularity of Sun Tzu in Western military discourse is underpinned by the view that a strong Russia (including the former Soviet Union) and/or a strong China poses a huge challenge and threat to the US and its allies. This is testified to by the fact that, among 410 military texts analyzed in Corpus 2, China was the most frequently discussed nation, followed by the US and Russia (see Table 6.1). Furthermore, a large portion of the military texts quoting Sun Tzu considered China and/or Russia (and the former Soviet Union) to be the major opponent in the global competition for power and control, and they were envisaging a military conflict between the US and China or the US and Russia (or Soviet Union).

In the 1960s when Griffith's translation was published, there was a common fear among Westerners that China's rise in military power would become a great threat to world peace. Hart (1963, vi) articulated such worries in his "Foreword" to Griffith's translation when he said that a fresh translation of Sun Tzu became "all the more important in view of the re-emergence of China, under Mao, as a great military power".

With China's economic and military capacities increasing drastically since the 1980s' with the country's reforming and opening up, the West's concerns about China's alleged threat have become more severe. Lt. Col. Gauthier (1999), in her monograph *China as Peer Competitor? Trends in Nuclear Weapons, Space, and Information Warfare*, wrote that there was a great potential for China to emerge as a competitor with the US in the coming decades. Gauthier examined three critical areas, namely nuclear weapons, space technology and information warfare, and asserted that US military advantages over China were narrowing. China had developed nuclear weapons increased accuracy, mobility and range. Its space program had made significant technological development in the military realm. The Chinese efforts in information warfare showed that China was preparing to wage asymmetric warfare against a more powerful adversary. Gauthier thought that China would become an adversary if Washington mishandled its relationship with Beijing.

The rise of China as a military power thanks to its modernization is also the topic of the monograph *China's Military Modernization: Building for Regional and Global Reach* by Richard Fisher (2008). Fisher argued that

Table 6.1 Nations most discussed in Corpus 2

No.	Nations	word list	Word frequency	Total
1	China	China	21274	36421
		Chinese	15147	
2	US	U.S. (US)	16125	26332
		USA	1024	
		America(n)	9183	
3	Soviet Union	U.S.S.R.	130	12297
	(Russia)	Soviet (Union)	7818	
		Russia(n)	4349	

"China's penchant for secrecy and deception stratagems, which was based on venerated historic treatises of statecraft such as that of Sun Zi", cloaked its military development (Fisher 2008, 5). Fisher listed a number of China's threatening foreign policy choices, including (1) preparing for a war against Taiwan, simultaneously risking war with the US, possibly resulting in long-term hostility between China and the West; and (2) pursuing political-military hegemony in Asia, specifically by trying to push out American influence (Fisher 2008, 1–5). In Fisher's view, China was making disturbing choices while extending its global influence, which would eventually challenge the existing status of the US as the only superpower:

> The Chinese Communist Party-led government is not satisfied with a world order in which the United States is the dominant power. While Chinese leaders acknowledge their growing dependence upon global good will for vital commercial and resource access, recent experience shows that CCP leaders will seize opportunities to alter power relationships and power balances. Their actions will very likely include the calculated but decisive use of military force.
>
> (Fisher 2008, 9)

In his research paper "Secrecy and Stratagem: Understanding Chinese Strategic Culture", which quotes the translated Sun Tzu frequently, Mahnken (2011) presented an appreciation of the main tenets of Chinese strategic culture. He also explained two compelling motives to study Chinese strategic culture: (1) the Chinese themselves see it as an important determinant of their behavior and that of others (Mahnken 2011, 2–3); and (2) to help the US to gain insight into Chinese decision-making. His paper was based on the perception that the US is in a military competition with China that will continue for the foreseeable future.

Similarly, McNeilly and McNeilly (2001, 4), whose monograph features six principles of modern strategy extracted from the translated Sun Tzu, pointed out that the understanding of the historical importance and modern relevance of *AOW* "becomes more critical" as China "moves to attain potential superpower status in the twenty-first century". In their view, China's huge population, increasing economic might, rapid technological development and military modernization make it a significant player in Asia, and one of the threats the US was facing since the breakdown of the world order after the Cold War (McNeilly and McNeilly 2001, 106). At the same time, they argued, China's leaders rely on strategic lessons from China's history, including those of Sun Tzu, to build their nation's strategy for the coming decades. Therefore, the West, particularly US diplomatic and military leaders, should learn as much as possible about China's strategic philosophy and how to best use Sun Tzu's principles. The knowledge of Sun Tzu's concepts should be systematically spread throughout both their military and diplomatic institutions. For the US and its Western allies, "it is not only necessary to accept and understand

Sun Tzu's strategic philosophy but also to analyze, compare, debate, and finally come to grips with the military theories of Sun Tzu versus those of Carl von Clausewitz" (McNeilly and McNeilly 2001, 197). They warned that while China is becoming more technologically advanced and militarily powerful and the world is becoming more uncertain, the US "must get its debt under control, it cannot afford to lose its technological edge and resulting military superiority" (McNeilly and McNeilly 2001, 205).

Ota (2014) also believed that contemporary Chinese strategy is heavily influenced by Sun Tzu, with its emphasis on deception and espionage. He spoke rather bluntly: "what should we do for countering Chinese strategy? We have to know and use Sun Tzu against China" (Ota 2014, 80). For instance, Ota suggested that Chinese moral influence and the legitimacy of its Communist Party could be damaged through news reporting about Chinese leaders' activities.

In numerous documents about contemporary US strategy, China is wrongly listed as a pressing threat to US national security and interests. *The US National Defense Strategy 2008* stated that, to disrupt American traditional advantages, China was developing technologies such as the anti-satellite capabilities and cyber warfare. It declared that the US Defense Department will "respond to China's expanding military power, and to the uncertainties over how it might be used, through shaping and hedging", and invest substantial "resources in ways that favor key enduring US advantages" (Department of Defense USA 2008, 10). In a report named "Revising US Grand Strategy Toward China" by the Council on Foreign Relations, it is advocated that the US should adopt "a new grand strategy toward China that centers on balancing the rise of Chinese power rather than continuing to assist its ascendancy" (Blackwill and Tellis 2015, 4), because China's rise thus "far has already bred geopolitical, military, economic, and ideological challenges to US power, US allies, and the US-dominated international order" (Blackwill and Tellis 2015, 5). The report suggests some measures to accomplish the shift to a process placing less emphasis on support and cooperation and more on economic and military pressure and competition, such as new trade arrangements in Asia excluding China, and more capable and more active US air and naval presence in the Asia-Pacific region.

The above investigation reveals that most US strategists cited and applied *Sun Tzu* with a purpose to renovate their own strategic thinking, to maintain their military supremacy and hegemonic power and to maintain the power inequality between the US and any other countries from which they may profit. Although the strategists have a good knowledge of Sun Tzu's peace-oriented principle of win without fighting, they turn blind eyes to China's constant efforts to avoid unjust wars, maintain peaceful co-existence and establish a Community of Shared Future for Mankind. These strategists are still confined both by their own strategic bias that unless the US is the only military power around the globe there is no security for them and by the Thucydides Trap. Their minds are blinded and plagued by ideological bias and stereotypes, which will be discussed in the next section.

6.3.2 *Ideological and cultural disparity*

Translation activities can never escape the grip of ideology. This is also true of the translation and re-canonization of the *AOW* in the West. In the following section, the ideological constraints over the translation and reception of Sun Tzu in the West will be discussed in detail.

When Giles was translating *AOW*, the prevailing ideology in the West was marked by ethnocentrism and the fear of the Yellow Peril, in particular. Giles' respect for ancient Chinese military culture, which is evident in his annotated translation, manifested his resistance to the mainstream ideology against oriental culture in his time. However, such respect and his stress on pacifism regretfully went almost unnoticed when his translations were received in the West.

Ethnocentrism was a mainstream conception in the West in Giles' age. For instance, as Giles (Giles 1907, 10) pointed out, Confucius' philosophy was voted boring, commonplace, shallow, disjointed and unsatisfying in Europe. Meanwhile, Confucius "was blamed for his materialistic bias, for his rigid formalism, for his poverty of ideas, for his lack of spiritual elevation" (Giles 1907, 10). This was due to the fact that "comparisons, much in his disfavour, were drawn between him and the founders of other world-systems of religion and ethics" (Giles 1907, 10). Ethnocentrism was also evident in the translations of Chinese cultural classics for the English audience. In the introduction to his translation of *The Sayings of Confucius*, Giles noted that two forerunning translators of *Analects* harbored an ethnocentric attitude toward Confucianism and the great sage Confucius. Specifically, James Ledge's translation, though an outstanding monument of research and erudition, revealed his preconceived bias that Confucianism must "at every point prove inferior to Christianity" (Giles 1907, 12). Another translator, William Jennings, held a prejudice that led him to view Confucius as non-religious, selfish and a poor parent.

Giles refuted these discriminatory assessments of Confucius with a detailed narration of Confucius' life and by setting the ancient Chinese social-cultural context. He announced that "Confucius was the prince of philosophers, the wisest and most consummate of sages, the loftiest moralist, the most subtle and penetrating intellect that the world had ever seen" and that Confucius' sayings are "the very epitome of wisdom" (Giles 1907, 8,10). Giles' opposition to ethnocentrism is rather obvious. Similarly, when translating *The Art of War*, Giles re-established a respectable image of Sun Tzu and his esteemed Chinese military cultural identity. He transcended ethnocentrism in his presentation of the cultural other.

To be specific, what Giles challenged in his translation of *AOW* is the fear of the Yellow Peril (sometimes called Yellow Terror). The Yellow Peril, a negative stereotype drawn from the ethnocentric view of Chinese immigrants (Lee and Kim 2014, 240) and other Asian people, was a widely used term and a popular theme in Western literature and journalism in the late nineteenth and early twentieth centuries. It portrayed an imagined threat from swarms of hardworking yellow-skinned immigrant Chinese to white people's living standards and cultural values. It also referred to the irrational panic and belief that China, with a population of 400 million, would launch wars with a huge, well-equipped army against Western countries and eventually wipe out Western civilization (Prince 1987, 2). Although China was defeated twice in the Opium Wars launched by the British army to impose the opium trade, the West as a whole still conjured up the Chinese military menace, which was widely accepted believed in Britain. The 1900 Boxer Rebellion in Beijing was an uprising against the spread of Western influence. Western diplomats, foreign civilians and soldiers were placed under siege for 55 days and some were killed by Boxers. Later, The Eight-Nation Alliance brought 20,000 armed troops to China, lifted the siege of the Legations and in revenge plundered Peking. The Boxer Rebellion, however, reinforced the Western fear of a military Yellow Peril (Frayling 2014; Prince 1987, 50). Paradoxically, the British colonists again "managed to cast the oppressed victims as a threatening, expansionist foe" (French 2014).

Giles disagreed with this ungrounded fear of Chinese military menace. He criticized the Western troop's outrageous acts in Peking after the Boxer Rebellion was put down. Giles (1910, 119) pointed out in his note that the Chinese commentators' injunction not to rape and loot "may well cause us to blush for the Christian armies that entered Peking in 1900 AD". He also managed to reshape for the reader an image of the best military treatise with a strong inclination for peace, as well as an image of a peace-loving nation.

Despite Giles' efforts in conveying the message of peace in his translated Sun Tzu and his challenge of the Western fear of Yellow Peril, most Western strategists who studied and applied the translated Sun Tzu chose to neglect the peace orientation. Instead of moderating their belligerence, they opted to view *AOW* as a guide to help them keep their advantages in various battlefields, to counter and defeat China, the Soviet Union and North Korea.

Unlike Giles, Griffith, as a translator and strategist from the US military sphere, was strongly influenced by the dominant anti-communist ideology of his time: the Cold War mentality. He didn't see much about a peace orientation in Sun Tzu, as Giles did. Instead, he firmly believed that his translation was a tool to understand the strategies used by communist officers and a device to defeat communism.

In a note to his translation of Mao Tse-tung's *On Guerrilla Warfare*, Griffith (1943) wrote that the influence of Sun Tzu on Mao's military thought was apparent. The tactics of Sun Tzu were in large measure the tactics of China's guerrillas. More importantly, Mao Tse-tung was a leader of the Chinese Communist Party and the former political commissar of the Fourth Red Army, who proposed that unlimited guerrilla warfare with vast time and space factors can establish a new military process when thoroughly organized from the military, political and economic point of view. Thus, Griffith found what Mao had written of this new type of war important, because up until then the Marine Corps had only encountered relatively primitive and strictly limited guerrilla war.

In another journal paper, "Some Chinese Thoughts on War", Griffith (1961) summed up some principles of the Chinese communist army on the basis of ancient strategists including Sun Tzu. His intention in doing this was firmly rooted in the confrontation between communism and capitalism, East and West. He explained that: "one of the most challenging problems now confronting western intelligence organizations is to acquire fundamental information on which to base reasonably realistic estimates of Red Chinese military capabilities" (Griffith 1961, 40). To better understand the military developments and strategic thought on the Chinese mainland, he wrote, the published "works of leading Communists must also be consulted" and the doctrine of the Chinese army should not be neglected (Griffith 1961, 40). In Griffith's view, since the sources of the current doctrine were traceable to remote antiquity, it was important to translate the works of Sun Tzu and Wu Ch'i.

In a book review of Griffith's translation of Sun Tzu, Boorman and Boorman (1964) remarked that Griffith's translation was significant because Sun Tzu's precepts were heeded both by China's Nationalist and Communist leaders in the twentieth century. In the Nationalist camp, "Chiang Kai-shek has long been known as a student and avid collector of editions of Sun Tzu" (Boorman and Boorman 1964, 132). The principal military theorists, Chiang Fang-chen (1882–1938) and Yang Chieh (1889–1949), both produced commentaries on *AOW*. However, the influence of Sun Tzu on the Chinese Communists remained somewhat unknown though such influence may have been considerable. Aware of this, Griffith included in his introduction an estimate of Sun Tzu's significance and established the links between Sun Tzu's theory and Mao's strategy, the "remarkable similarity" between some passages in Sun Tzu and the well-known set of Red Army slogans. Griffith's handling of such paratexts was grounded in the Cold War mentality. As Hamilton (1998, 17) pointed out: for many strategists and scholars, "the study of works like Sun Tzu's was viewed as vital in helping to solve the puzzle of communist uprisings and subversive political successes throughout Asia".

In addition to Griffith, other Western strategists have also pointed to the fact that deeply rooted behind the translation and re-canonization of Sun Tzu is the ideological disparity between East and West. We assumed that the prevailing ideology in the military texts quoting Sun Tzu could be identified by counting the occurrences of ideological terms. A survey was made of the key terms that may strongly express or indicate ideology in Corpus 2's 410 texts. The result of the survey is shown in Table 6.2 and reveals that "communist (communism)" is the most frequently used ideological term in the texts, with an average occurrence rate of seven times per text.

To further understand the main themes in discussions about communism in Corpus 2, we firstly came up with a list of word clusters beginning with "communist". Clusters with the same/similar meaning but in different forms

Table 6.2 Ideological terms in Corpus 2

No.	Ideological terms	word list	word frequency	Total
1	Communism and Socialism	Communism	368	3065
		Communist	2470	
		Socialism	78	
		Socialist	149	
2	Capitalism	Capitalism	92	174
		Capitalist	82	

Table 6.3 Word clusters beginning with "communist" in Corpus 2

Rank	Word Cluster	Frequency
1	communist party/parties	488
2	communist army/armies/force(s) /soldiers/troops/military	193
3	communist bloc/countries/world/ regime(s)/state(s)	164
4	communist leader(s)/leadership	85
5	communist China/Chinese	77
6	communist aggression/conquest/insurgency/control/ expansion	96
7	communist revolution(ary)	65
8	communist movement	36
9	communist control/rule	36
10	communist ideology	26
Total		1266

were sorted out. For instance, the word clusters "communist party" and "communist parties" were grouped together, as were "communist army" and "communist force(s)". The most frequent 10 clusters are listed in Appendix 6. The survey found that the most frequent word cluster is "communist party (parties)", the second most was "communist armies", and the third was "communist bloc/countries" (see Table 6.3). That the ideological disparity between communism and capitalism greatly conditions *AOW*'s translation and its reception in the West is confirmed by the high frequency of the word "communism(t)" and its word clusters.

Beginning in the 1990s, enthusiasm about Sun Tzu was spurred by US exaggeration of China's threat to US national security and the advocacy for a competitive edge for strategic advantage. China Threat Theory is deeply rooted in the distrust and conflict between capitalism and communism, specifically the alleged conflict between US-style democracy and the different form of governance in China.

As Hooker (2014) alleged, US now counters the rise of China with its "Rebalance to Asia" directives and opposes Chinese territorial moves in the

East and South China Seas; both of which are consistent with longstanding US grand strategy. It is further stated that:

> A peaceful, nonhostile peer nation or grouping of nations (such as the European Union) poses no strategic threat to the United States. An authoritarian great power, possessed of both military and economic means and an apparent desire to enlarge and expand them, could in time pose a direct, existential threat to American national security.
>
> (Hooker 2014, 16)

Hooker's opinion explains the ideological reason why an economic powerful Europe is not considered to be a strategic threat to the US, while the socialism practiced in China with its different form of governance is seen to be at odds with US capitalism and its so-called standard style of democracy. And this ideological disparity has become a deep-rooted impetus for Western military strategists to translate *AOW*, study it and apply it.

Sun Tzu's reception in the West has also been influenced by the West's centricity, arrogance and cultural bias against non-Western values. "Rather than defending itself against real threats, the West has been possessed by fantasies about projecting its values throughout the world. When these values are derided and rejected by leading sections of the West itself, it is a vain enterprise" (Gray 2020, 41). Sun Tzu attached great importance to the knowledge of not only the enemy but also of yourself, including your limitations. However, clouded by ideologies that accord them a privileged place in history, "Western elites seem resistant to self-knowledge". How much the West can learn from *AOW*, a text of matchless wisdom, "is another matter" (Gray 2020, 41).

The war in Afghanistan reflected the fact that "Western culture holds a self-interested perspective that American forces 'have the right to retribution' through offensive military action against the Taliban and Al Qaeda due to the tragedy of 9–11" (Zweibelson 2010, 2). For Western society, the Afghan War was viewed as justified through a Western political and social lens, but "it translates poorly in the Islamic World and specifically Afghanistan" (Zweibelson 2010, 2). Blinded by Islamic ideology and having suffered from a history of war and occupation by many imperialistic powers, a majority of the Afghan people is illiterate, living in tribal-based societies in geographically remote areas and inhospitable to outsiders. While Western society portrayed American soldiers deployed in Afghanistan as brave liberators or saviors, "the general Islamic population sees the same soldiers as infidel occupiers forcing tribal groups into an unnatural national entity that conforms to western societal values" (Zweibelson 2010, 2).

Furthermore, the uses of Sun Tzu in Western military discourse are greatly constrained by the differences between Eastern and Western strategic culture. Although US strategists successfully ushered some of Sun Tzu's ideas into Western strategic thinking, they still blundered as they disobeyed Sun Tzu's rules. Their negligence and mistakes in applying Sun Tzu are evident in the following two ways, in which the huge drag of Western strategic heritage is visible.

To begin with, most strategists could not escape the bondage of Clausewitzian traditional principles, even though some agreed that, in general, "Sun Tzu instructed on war but taught peace", viewed a justified and unavoidable war as the last resort and called for transformation but not annihilation of the enemy (Hwang and Ling 2009, 71). Griffith recognized that Sun Tzu "did not conceive the object of military action to be the annihilation of the enemy's army, the destruction of his cities, and the wastage of his countryside". However, when the US launched its invasions of Iraq and Afghanistan, they abandoned Sun Tzu's tenets and turned to their own military tradition. When discussing the US military operation in Afghanistan, Major Ben Zweibelson, who participated in Operation Iraqi Freedom, pointed out that:

> Over the past nine years United States counterinsurgency strategy reflected a reliance on Clausewitzian industrial-era tenets with a faulty emphasis on superior western technology, doctrine fixated on lethal operations, and a western skewed perspective on *jus ad bellum* [just cause for war]. American military culture is largely responsible for the first two contextual biases, while western society is liable for the third in response to September 11, 2001.
>
> (Zweibelson 2010, 1)

The mainstream Western military strategy is culturally different from Sun Tzu's ancient military philosophy. Western dependance on supremacy gained by advanced technology and overwhelming firepower has made the American counterinsurgency doctrine largely fatal in nature and enemy focused. Under Clausewitz's instruction, generations of Western military officers have believed that the annihilation of the enemy's army is the chief goal in all combat. As a result, all political-military conflict has resulted in offensive actions in which the attrition of the enemy forces were a universal requirement (Zweibelson 2010, 1). To address the operational failure in Afghanistan, the U.S. military "should acknowledge the errors of the past nine years of over-reliance on failed counterinsurgency strategy" (Zweibelson 2010, 3), and should replace the teachings of Carl von Clausewitz with those of Sun Tzu, in favor of more appropriate counterinsurgency alternatives (Zweibelson 2010, 1).

Secondly, the Western incorporation of Sun Tzu has remained at the level of tactics rather than of strategy. Sun Tzu declared: "strategy without tactics is the slow road to victory" but "tactics without strategy is the noise before defeat". This means that if a state makes a wrong decision on strategy, no matter how excellent its tactics are, it can never save a war fought under a ruined strategy. "In Afghanistan, American strategy flunked Sun Tzu" (Gentile 2012). From the start of the war in 2001, America had a quite limited core policy goal focused on disrupting, disabling and eventually defeating al

Qaeda. However, to achieve this limited goal, the US attempted armed nation-building from the very beginning, investing huge amounts of blood and funds in Afghanistan, and it did not work. As early as in 2012, Gentile (2012) reported that American-style counterinsurgency has failed in Iraq and it was currently failing in Afghanistan. In a similar view, Gray (2020, 61) predicted shortly before US withdrawal from Afghanistan: "the nearly 20-year presence of Western forces in Afghanistan has served no discernible purpose", and it would become another strategic failure.

Similarly, in Iraq, the "lessons of Sun Tzu for the West are discomforting" (Gray 2020). Using a pretext of disinformation, the US launched and led the Iraq War that had no definable strategic objectives. As some predicted at that time, the effect of overthrowing Saddam's rule was to ruin the country of Iraq, and to set free as well as irritate Islamist fundamentalism. The West coalition imposed regime change in Libya in 2011, but it brought about bigger disasters, putting the fate of the country into the hands of jihadists and people smugglers, and triggering a costly and devastating civil war.

6.4 Summary

The analysis in this chapter has revealed the fact that the translation and re-canonization of *AOW* in Western military discourse has been constrained by factors such as situational context, including the translator's social identity and target readership, the need for strategic innovation in Western military institutions, the competition for power and the prevailing ideological disparity.

Giles' social identity as an erudite sinologist with an admiration for Chinese culture and his target readers—including his brothers, who were British soldiers, and those interested in Chinese culture—propelled him to add commendatory remarks in his paratexts and stress the peace orientation in *AOW*. Griffith was an American general and a military expert on Chinese strategists (especially Mao Tse-tung), and his target readership included high-ranking American strategy decision makers as well as military staff. These factors determined that the emphasis in Griffith's translation is on the use of strategic concepts and principles to achieve victory in conflicts. They also led Griffith to attempt to modernize and globalize Sun Tzu in the paratexts for contemporary readers.

The reception of the translated Sun Tzu in Western military discourse is affected first of all by the need for strategic innovation. Western strategy has long been monopolized by Clausewitz's *On War*. However, the huge casualties in two world wars, the risk of human extinction brought about by the nuclear arms race, the US failures in the Korean War and Vietnam War have compelled Western military theorists to seek alternative solutions and eventually they turned to Sun Tzu for inspiration. In the twenty-first century, severe challenges from unconventional and new warfare, such as global terrorism, asymmetrical war and cyber war, have made the strategic incorporation between the East and West more urgent.

The translation and reception of Sun Tzu in Western discourses are also constrained by the ideological cultural disparity. When Giles was translating, the West was biased due to its fear of the Yellow Peril. In defiance of that prevailing ideology, Giles tried to establish a pacifist image of Sun Tzu, though his efforts went unheeded. Griffith, driven by the confrontation between the capitalist and communist bloc in the Cold War, viewed his translation as a weapon firing back at the oriental nations where communism is believed and socialism is practiced.

Like Griffith, most Western strategists, though having attempted to learn from Sun Tzu, failed to break away from the Cold War mentality that led them care for nothing but gaining supremacy over communism. The reception of Sun Tzu also plagued by the China Threat theory, which resulted from the ideological disparity between the US and China and in turn dragged the US into its attempts to seize military power in the East Asia area. Sun Tzu's reception in the West was also influenced by Western centricity, arrogance and cultural bias against non-Western values. This was exemplified by America's earnest efforts to force democracy in Afghanistan and its failure to understand the tradition, culture and history of the Afghani people. In particularly, the reception of *AOW* has been constrained by the differences between the Eastern and Western strategic cultures and specifically crippled by the West's over-reliance on the Clausewitzian tradition. Misled by this Western strategic culture, American strategists have placed a faulty emphasis on superior military technology, lethal operations and a preference for tactics over strategy, which led to their failures in military operations in Libya, Iraq and Afghanistan.

References

Atkinson, Carol. 2018. "Hybrid Warfare and Societal Resilience: Implications for Democratic Governance." *Information & Security: An International Journal* 39 (1): 63–76.

Bartley, Caleb M. 2005. "The Art of Terrorism: What Sun Tzu Can Teach Us About International Terrorism." *Comparative Strategy* 24 (3): 237–251.

Bingöl, Oktay. 2017. "Hybrid War and Its Strategic Implications to Turkey." *Gazi Akademik Bakis Dergisi* 11 (21): 107–132.

Blackwill, Robert D., and Ashley J. Tellis. 2015. *Revising US Grand Strategy toward China. Council Special Report No. 72*. New York: Council on Foreign Relations.

Boorman, Scott A., and Howard L. Boorman. 1964. "Mao Tse-tung and the Art of War." *The Journal of Asian Studies* 24 (1): 129–137.

Buffaloe, David L. 2006. *Defining Asymmetric Warfare. The Land Warfare Papers*. Arlington, VA: Institute of Land Warfare, Association of the United States Army.

Coker, Christopher. 2003. "What Would Sun Tzu Say About the War on Terrorism?" *RUSI Journal* 148 (1): 16–19.

Delisle, Jean. 1999. *Translation Terminology*. Amsterdam: John Benjamin's Publishing Company.

Department of Defense USA. 2008. *National Defense Strategy*. Washington, DC.

Fisher, Richard D. 2008. *China's Military Modernization: Building for Regional and Global Reach*. Westport, CT: Praeger Secruity International.

Frayling, Christopher. 2014. *The Yellow Peril: Dr Fu Manchu & the Rise of Chinaphobia*. London: Thames & Hudson.

French, Philip. 2014. "The Yellow Peril: Dr Fu Manchu & the Rise of Chinaphobia (book review)." *The Guardian*, October 20. www.theguardian.com/books/2014/oct/20/yellow-peril-fu-manchu-rise-chinaphobia-review-factors-fear-china (Accessed September 10, 2015).

Gauthier, Kathryn L. 1999. *China as Peer Competitor? Trends in Nuclear Weapons, Space, and Information Warfare. Maxwell Paper No. 18*. Maxwell Air Force Base, AL: Air War College.

Geers, Kenneth. 2011. *Sun Tzu and Cyber War*. Tallinn, Estonia: NATO Cooperative Cyber Defence Centre of Excellence.

Gentile, Gian. 2012. "American Strategy in Afghanistan Flunks Sun Tzu." Jerusalem Post. July 3. www.jpost.com/Opinion/Op-Ed-Contributors/American-strategy-in-Afghanistan-flunks-Sun-Tzu (2012, May 20).

Giles, Lionel. 1907. *The Sayings of Confucius: A New Translation of the Greater Part of the Confucian Analects (with Introduction and Notes by Lionel Giles)*. London: John Murray.

Giles, Lionel. 1910. *On the Art of War: The Oldest Military Treatise in the World (Translated from the Chinese with Introduction and Critical Notes by Lionel Giles)*. London: Luzac.

Gray, Colin S. 1999. *Modern Strategy*. Oxford: Oxford University Press.

Gray, John. 2020. "What Sun Tzu Knew." *New Statesman* (January-February): 38–41. www.newstatesman.com/sun-tzu-the-art-war-politics (Accessed 18 May, 2020).

Greathouse, Craig B. 2013. "Cyber War and Strategic Thought: Do the Classic Theorists Still Matter?" In *Cyberspace and International Relations: Theory, Prospects and Challenges*, eds. Jan-Frederik Kremer and Benedikt Müller, 21–40. New York: Springer.

Griffith, Samuel B. 1938. "North China, 1937." *Marine Corps Gazette* 22 (4): 23–24, 46–53.

Griffith, Samuel B. 1941. "Guerrilla Warfare in China." *Marine Corps Gazette* 25 (2): 19–50.

Griffith, Samuel B. 1943. "That Man Suntzu." *Marine Corps Gazette* 27 (4): 3–6.

Griffith, Samuel B. 1961. "Some Chinese Thoughts on War." 45 (4): 40–44.

Griffith, Samuel B. 1963. *The Battle for Guadalcanal*. Annapolis: Nautical & Aviation.

Griffith, Samuel B. 1964. "The Glorious Military Thought of Comrade Mao Tse-Tung." *Foreign Affairs* 42 (4): 669.

Griffith, Samuel B. 1965. "Communist China's Capacity to Make War." *Foreign Affairs* 43 (2): 217.

Griffith, Samuel B. 1966. *Peking and People's Wars: An Analysis of Statements by Official Spokesmen of the Chinese Communist Party on the Subject of Revolutionary Strategy*. New York: Praeger.

Griffith, Samuel B. 1967. *The Chinese People's Liberation Army*. New York: McGraw-Hill Book Company.

Griffith, Samuel B. 1969. "Chinese Capabilities and Strategies for the 1970's." *Marine Corps Gazette* 53 (8): 48–49.

Griffith, Samuel B. 1974. *History of the Second World War*. Hicksville, NY: BPC Publishing.

Griffith, Samuel B. 1976. *The War for American Independence: From 1760 to the Surrender at Yorktown in 1781.* New York: Doubleday.

Hamilton, Donald W. 1998. *The Art of Insurgency: American Military Policy and the Failure of Strategy in Southeast Asia (Foreword by Cecil B. Currey).* Westport, CT and London: Praeger.

Handel, Michael I. 2001. *Masters of War: Classical Strategic Thought.* 3rd ed. London and Portland: Frank Cass.

Hart, Liddell. 1963. "Foreword." In *Sun Tzu: The Art of War*, trans. Samuel B. Griffith, vi–vii. New York: Oxford University Press.

Homayoun, Assad. 2004. "Sun Tzu: The Newest View." *Defense & Foreign Affairs Strategic Policy* 32 (10): 9.

Hooker, Richard D. 2014. *The Grand Strategy of United States.* Washington D.C.: National Defense University.

Hwang, Ching-Chane, and L. H. M. Ling. 2009. "The 'Kitsch' of War: Misappropriating Sun Tzu for an American Imperial Hypermasculinity." In *Gender and Global Politics in the Asia-Pacific*, eds. Bina D'Costa and Katrina Lee-Koo, 59–76. Basingstoke: Palgrave Macmillan.

Johnston, Alastair I. 1995. *Cultural Realism: Strategic Culture and Grand Strategy in Ming China. Princeton Studies in International History and Politics.* Princeton, NJ and Chichester: Princeton University Press.

Johnston, Alastair I. 1999. "Sun Zi Studies in the United States." (unpublished paper) Cambridge, MA: Harvard University.

Lee, H. C., and K. Kim. 2014. "Ethnocentrism." In *Race and racism in the United States: An Encyclopedia of the American Mosaic*, eds. Charles A. Gallagher and Cameron D. Lippard, 240–241. Santa Barbara CA: Greenwood.

Lunas, Frederic W. 2011. *The Modern Application of Sun Tzu's Art of War: Improving Application to Cyber.* Wright-Patterson Air Force Base, OH: Center for Cyber Research (US).

Mahnken, Thomas G. 2011. *Secrecy and Stratagem: Understanding Chinese Strategic Culture.* Double Bay, NSW: Lowy Institute for International Policy.

Manko, Oleg, and Yurii Mikhieiev. 2018. "Defining the Concept of 'Hybrid Warfare' Based on the Analysis of Russia's Aggression against Ukraine." *Information & Security: An International Journal* 41: 11–20.

Marshall, Andrew S. 1999. "Is the Sun Tsu's Art of War Relevant to the Modern Air Force?". *RUSI Journal* 144 (5): 57–60.

McNeilly, Mark, and Mark R. McNeilly. 2001. *Sun Tzu and the Art of Modern Warfare.* New York: Oxford University Press.

Miller, Matthew K. 2001. *Sun Tzu and the Art of (Cyber) War: Ancient Advice for Developing an Information Security Program.* Boston: SANS Institute.

Minford, John. 2008. "Forward. In *Sun Tzu The Art of War* (Bilingual Edition with Complete Chinese and English Text), trans. Lionel Giles. Tokyo, Rutland, VA and Singapore: Tuttle Publishing.

Monaghan, Sean. 2019. "Countering Hybrid Warfare: So What for the Future Joint Force?" *Prism: A Journal of the Center for Complex Operations* 8 (2): 82–98.

Ota, Fumio. 2014. "Sun Tzu in Contemporary Chinese Strategy." *Joint Force Quarterly* 73: 76–80.

Prince, Graham. 1987. "The Yellow Peril in Britain 1890–1920." MA Thesis. McGill University.

Rice, Sean P. 2006. "Sun Tzu: Ancient Theories for a Strategy Against Islamic Extremism." Research paper. US Army War College.

Richter, Paul. 1991. "Ancient Doctrine Guiding Futuristic Warfare in Gulf: Strategy: Sun Tzu's Tiny Book, Written More Than 2,500 Years Ago, Is Influencing U.S. and Iraqi Tactics." February 18. www.latimes.com/archives/la-xpm-1991-02-18-mn-1191-story.html (Accessed June 21, 2020).

Snyder, Jack. 1990. "The Concept of Strategic Culture: Caveat Emptor." In *Strategic Power: USA/USSR*, ed. Carl G. Jacobsen, 3–9. London: Palgrave Macmillan UK.

Snyder, Jack L. 1977. *The Soviet Strategic Culture: Implications for Limited Nuclear Operations, RAND R-2154-AF*. Santa Monica, CA: The Rand Corporation.

Stuart, Douglas T., and William T. Tow. 1981. "The Theory and Practice of Chinese Military Deception." In *Strategic Military Deception: Pergamon Policy Studies on Security Affairs*, eds. Donald C. Daniel and Katherine L. Herbig, 292–316. New York: Pergamon Press Inc.

Szymanski, Paul. 2020. "Techniques for Great Power Space War." *Strategic Studies Quarterly* 13 (4): 78–104.

Vetter, Lawrence C. 1996. *Never without Heroes: Marine Third Reconnaissance Battalion in Vietnam, 1965–70*. 1st ed. Ivy Books Vietnam/Non-fiction. New York: Ivy Books.

Williams, Thomas J. 2003. "Strategic Leader Readiness and Competencies for Asymmetric Warfare." *Parameters* 33 (2): 19–35.

Yuen, Derek M. C. 2014. *Deciphering Sun Tzu: How to Read The Art of War*. New York: Oxford University Press.

Zaman, Rashed U. 2009. "Strategic Culture: A "Cultural" Understanding of War." *Comparative Strategy* 28 (1): 68–88 (Accessed July 30, 2014).

Zweibelson, Ben. 2010. "The US in Afghanistan: Follow Sun Tzu Rather Than Clausewitz to Victory." *Small Wars Journal* (December): 1–6.

7 Conclusion
Findings, implications and beyond

In the previous chapters, we have conducted a textual analysis of the translations of Sun Tzu, a discourse analysis of the reception of *The Art of War* in the West, and a social-cultural analysis of the constraints on the translation and reception of Sun Tzu. This chapter draws a conclusion from the above-mentioned investigation. It summarizes the findings and their implications, reflects on the limitations of the current research and outlines some potential perspectives for future research.

7.1 Summary of the findings

As a most intelligent general and a great writer, Sun Tzu noted down his acute perception of war with a large number of highly condensed military terms, strategic principles and metaphoric expressions, among them the peace-oriented principle of victory without fighting is paramount. Sun Tzu's systematic inquiry into strategy boasts a large army of commentators and students in China, each adding some new pages and wise reflections to Sun Tzu's writing throughout the ages. As a result, *The Art of War* gradually ascended to the throne of the most influential military classic in China and became representative of the core values of Chinese strategic culture. The multiple paratexts added by Sun Tzu's followers such as comments, prefaces and introductions have enriched the content of the core text and embellished on Sun Tzu's strategic thinking, adding to the persuasive power of the text and venerating of Sun Tzu.

In 1905, *The Art of War* began its entry into the Western military discourse with its English translations. Among more than 60 translators who issued numerous editions and reprints, Lionel Giles and Samuel Griffith are two key English translations. Our analysis into their translations has shown that both of them feature a close collaboration between the core text and paratexts which contributes to the effective reconstruction of ancient Chinese strategic culture.

In both Giles' and Griffith's translations of Sun Tzu's core text, the domestication approach is much more frequently used than the foreignization approach in rendering military concepts and principles. Consequently,

DOI: 10.4324/9781003025726-7

intelligibility and fluency are achieved while the profundity of the terms and principles as well as the exotic flavor of ancient Chinese military culture is sacrificed. Fortunately, paratexts, particularly translators' notes, come to aid the translation by compensating for the deficiency of core texts. They make up for the exotic flavor by providing foreignizing transliteration, literal translation and in some cases Chinese ideographic letters. In the few cases where Sun Tzu's military terms and principles are foreignized in the core texts, paraphrasing is used in the notes to enhance intelligibility. Thus, a delicate balance of domestication and foreignization is achieved by the concerted efforts of the core text and paratexts.

Paratexts in these translations contribute to the effective reconstruction of Chinese strategic culture with four approaches: supplementation, recontextualization, comparison and evaluation. To begin with, paratexts in both translations are furnished in large quantity to explain Sun Tzu's strategic thinking, to narrate the life story of Sun Tzu, to introduce the different versions of *The Art of War*, to describe its social settings and to reveal its impact. Consequently, the depth and width of ancient Chinese strategic culture are greatly enhanced. Secondly, paratexts in both translations are added to compare Chinese strategic culture with its Western counterpart. In Giles' translation, paratexts are elicited to compare Sun Tzu with a number of Western military generals to reveal the fact that Sun Tzu's theory has been testified to be true. In Griffith's translation, however, paratexts mainly contrast *The Art of War* with Clausewitz's *On War* to reach the conclusion that Sun Tzu's military thesis is more realistic, moderate, clear, profound and fresh. Comparison in effect brings the truth of Sun Tzu's strategic thinking into spotlight. Thirdly, commendatory evaluative remarks are added in the paratexts to build up respect toward Sun Tzu and to set up a positive image of ancient Chinese strategic culture. Giles himself acclaimed Sun Tzu's strategic thinking in his notes; while Griffith, together with the prestigious strategist Liddell Hart, applauded Sun Tzu's work. The positive evaluations are helpful in triggering interest and eliciting reception among readers. Fourthly, paratexts in Griffith's translation are employed to recontextualize ancient Chinese strategic culture. Sun Tzu is reincarnated as a modern general well-known around the globe. Sun Tzu's strategic principles versed in ancient language are enunciated through modern military terms. By recontextualization, the gap between the past and present, the East and West is bridged and an easy access to the translated strategic culture is offered.

Our discourse analysis, based on two corpora built up with over 490 texts quoting Sun Tzu, finds out that translations of Sun Tzu have been received with an increasing popularity in the West and consequently has had a profound influence upon the Western strategic culture. The positive reception of Sun Tzu was confirmed by the intensive direct and indirect quotations among a large sum of different types of military texts in the West. Not surprisingly, paratexts from translations were also quoted. The reception of the translated Sun Tzu has gone through three stages in Western discourse: at the individual, institutional and societal levels, to reach a status of re-canonization in Western strategic culture.

At the individual level, translations of *The Art of War* were interpreted, applied and developed by many strategic theorists, military officers and politicians like Liddell Hart, Michael Handel, Richard Nixon and John Boyd. They have made their contribution to the dissemination of Sun Tzu among many other individuals and pushed it toward the next stage. At the institutional level, the translated Sun Tzu is quoted extensively in the doctrines of the US Army, Navy, Air Force and Marine Corps. It has also been taught in professional military educational institutions as an important component of their curriculums. More importantly, Sun Tzu presided over the planning and execution of successful military operations including the Gulf War and the Iraq War. The influence of the translated Sun Tzu continued to expand and has reached beyond the military sphere. At the societal level, Sun Tzu is extensively cited in non-military discourse, such as in business and management, sports and entertainment. The popularity of Sun Tzu among non-military communities evidenced that *The Art of War* has also secured a canonical status in Western culture.

This study confirms that in addition to the translated core text, paratexts can have a direct and profound impact on the reception of a classic in the target culture. Giles' literal translation of the water simile led to its resonance with Liddell Hart. The paraphrasing of 奇正 (*ch'i* and *cheng*) in the core text inspired Hart's strategic theory of indirect approach, while the transliteration of these terms in paratexts were adopted by other strategists like John Boyd. Sun Tzu's life story translated from the SL paratexts contributed to the Shock and Awe Theory, which was later put into test in the Iraq War. The preliminary comparisons between Sun Tzu and Western strategists in paratexts blazed the trail for in-depth comparisons by later military theorists.

According to our social-cultural analysis, the translation and reception of *The Art of War* was closely constrained by a web of situational, institutional and social-cultural factors. Firstly, both translations were influenced by the translators' identity and their target readers. Giles' choice of translation approach and his treatment of the paratexts were directly influenced by his own identity as a sinologist and an admirer of Chinese culture as well as his target readership, which included his brothers, other British soldiers and those interested in Chinese culture. As for Griffith, a US marine officer and an expert on the military thoughts of Mao Tse-tung, his task of translation was to offer his military colleagues an inspiration for finding the solution to contemporary strategic issues. His target readers were high-ranking military officers and politicians. Griffith's emphasis, therefore, was placed on the strategic theory of Sun Tzu and its modern implications.

Secondly, the translation and reception of *The Art of War* in the West was also conditioned by the Western militaries need for strategic innovation. A series of global events that have occurred from the twentieth century till the present all called for strategic evolution: the Clausewitz-directed two world

wars resulting in massive casualties, the Cold War shrouded by the confrontation between communist and capitalist camps, the nuclear crisis risking the
destruction of the human race, as well as rampant global terrorism. Such
innovation needed fresh insights not only from within the Western strategic
tradition but also from outside. The translated Sun Tzu then served as a
source of inspiration for innovations in Western strategy.

Thirdly, we argue that the reception of translated Stun Tzu was also driven
by the competition and ideological disparity between the East and the West,
specifically between the US and China. On one hand, the US military institutions have used Sun Tzu as a means to achieve dominance over other nations
ever since the Second World War. With the military and economic rise of
China, the fear of the Chinese threat to US security has become stronger. The
US military believed that by knowing and using Sun Tzu, they could secure a
deep understanding of Chinese strategic culture, predict Chinese moves and
face the challenges of potential future military conflicts. On the other hand,
this ideological disparity is the impetus for Sun Tzu's introduction into the
West. In 1910, Giles tried to resist the prevailing ideology of his time, specifically the fear of the Yellow Peril, by emphasizing on the peace-orientation
in *The Art of War*. However, Giles' voice in favor of peace was seldom heard
afterwards. Unlike Giles, Griffith, like many other strategists, succumbed to
the alleged China Threat Theory. The popularity of Sun Tzu in the West is
inevitably underpinned by the conflict and hostility between communist and
capitalist beliefs.

Our findings have revealed that the translations of *The Art of War* have
exerted a profound influence on Western culture. With the combined efforts
of the core texts and paratexts, ancient Chinese strategic culture has been
effectively re-established, laying a solid foundation for the reception of Sun
Tzu. After having gone through different levels of reception, *The Art of War*
eventually establishes itself as a strategic classic in the target culture. Our
study shows that the translation of this military classic has led to innovation
in Western strategic thinking and brought about a huge impact on the way a
war is fought. It also played a part in shaping the target culture in military
and other spheres.

7.2 Implications and limitations

According to the above summary, we may sum up the implications of this
case study from four aspects: the importance of paratexts, the corpus-assisted
approach to the study of translation reception, the analytical framework for
classic translation and the impact of classic translation.

First of all, this study reveals the importance of the translator's choice
to use over paratexts. The findings suggest that translators and researchers,
especially those who deal with classics, need to become aware of the role
of paratexts in translations and to make the best use of them. Thanks to
their multiple forms, flexible lengths, varied authorships and more modes of

representation, paratexts can enrich the culture embodied in the core text, add to its persuasive power, and eventually become conducive to its reception into the target culture. Translators should not limit themselves within an either-or choice between the foreignization approach and the domestication approach in the core texts. Instead, they should go beyond to reach a set of tools that are offered by paratexts, such as supplementation, recontextualization, comparison and evaluation. To be specific, translators can selectively translate the original paratexts or supplement new ones to enrich the SL culture. They can recontextualize ancient cultures by putting them in a modern setting. They may compare the source culture with the target culture to inform the target readers of the difference and similarities between them. The translator's evaluation in paratexts, either commendatory or derogatory, may help establish an image of the source culture that the translator and readers desire. With multifunctional paratexts used in various combinations, translators can make up for limitations in the core text and better accomplish their task of source culture reconstruction.

The second implication mainly comes from the innovative use of quantitative investigations in study of translation and its reception. The findings suggest that a corpus-assisted approach is helpful in investigating issues, such as reconstructing a culture in translation, the reception of translated classics and the relevant social-cultural context analysis. With a comprehensive statistical investigation into the translating approaches for culture specific items, a full and clear picture about how a certain culture is translated can be mapped out. More importantly, quotations can be used to measure statistically the reception of translated texts in the target culture. With a collection of texts quoting translations, a corpus can be established to extract statistics according to certain parameters. A set of parameters might be useful, such as the number of texts quoting a translation, the average number of quotes in a text, the ratio between direct and indirect quotations and the negative and positive evaluative words with which the quotation is used. These tangible parameters can provide a panoramic view about how and to what extent a translated classic is received.

The third implication of this research is about the systematic assessment on the translation of a classic. As our case study of *The Art of War* has shown, the comprehensive framework to investigate translation, which includes textual analysis, discourse analysis and social-cultural analysis, is workable. Therefore, we argue that textuality, paratextuality, intertextuality and contextuality are the four elements for us to consider in an in-depth case study of translation. Textuality mainly involves the features of core texts, while paratextuality concerns the functions of paratexts. A discussion on the translation of a classic would be incomplete if it deals with the core text but excludes paratexts, and vice versa. Intertextuality, particularly in quotations, can be used as a tangible index to reflect on the reception of the translated texts. It can also be used as concrete evidence to measure the effect of translation approaches and translators' choices. Contextuality explains the situational and social-cultural constraints on the translation and reception of a given classic. It can help clarify the power relation and hidden ideology that push forward the translation of a certain classic and its dissemination in the target culture. These four integral and interrelated parts, namely textuality, paratextuality, intertextuality and contextuality, constitute a systematic framework for us to look into the issue of classic translation.

Last but not least, the study also showcases the profound impact of classic translation on the cross-cultural exchange. Translation of classics is an interlingual discursive event that is socially and culturally constituted and that in return constitutes society as well as culture. Like *The Art of War*, a classic work is usually representative of a certain culture in a community or a country. Its translation, in this sense, is more an issue of cross-cultural exchange than a matter of words. It shoulders the responsibility of reconstructing one culture for the target readers in a different culture. Translators need to take proper approaches and make wise choices in core texts and paratexts. With the source culture effectively reconstructed, a translated classic may exert its influence upon different entities in the foreign culture starting from individuals, to institutions and finally to the whole society. When the reception of the translated classic has gone through these stages with great popularity, it is safe to conclude that the canonical status of the translated source text is re-established in the target culture. We have to notice that the extent to which a translated culture is received in the target culture is constrained by a set of complicated factors including the target readers, institutional needs, power and ideology. More often than not, classic translations are constrained by power relations and underpinned by different attitudes, stances and ideologies, either private or public, positive or negative. The translated classic can be used as a tool to innovate the target culture, and at the same time, as a weapon against the community which hosts the source culture. Consequently, translation of classics becomes a rather complex activity of intercultural exchange, which needs meticulous study.

Although this study has its findings and implications, there are still some limitations to it. To begin with, this study in some way suffers from time constraints and limited access to more resources. Although we tried all means to collect as many texts currently available as possible while establishing the corpus of the translated Sun Tzu in the West, some texts from military institutions, especially those classified ones, are denied access and therefore they are not included in our study.

Secondly, as the textual analysis has focused on the translation of strategic culture, it is hardly possible to present a comprehensive and in-depth investigation of linguistic issues in the thesis. There is a need to probe deeper into more linguistic issues.

Thirdly, since the corpus mainly consists of texts from the US, this study focuses on how the target text impacts US strategic culture, and it would be more comprehensive if the research scope could have been expanded to cover more Western countries around the globe while investigating the reception of translated Sun Tzu.

Last but not least, although this book contains a comparative textual analysis of Giles' and Griffith's translations of *The Art of War*, it does not have the space to discuss other versions. Many other influential versions, such as Ralph Sawyers' and Roger Ames' translations, are also worthy of investigation.

7.3 Prospect for future work

With the reflection on the limitations of this study, we could continue future research on the translations of military classics by pursuing the following approaches. Firstly, in addition to the data drawn from the texts quoting the translations, other data sources can be exploited. For instance, we could launch a survey by means of questionnaires or interviews to look into the reception among current readers of *The Art of War,* especially soldiers and military officers. Since Sun Tzu is widely read and taught in many US military institutions, such valuable first-hand data would offer more insights about the translation and reception of Sun Tzu.

Secondly, the corpus about reception of the translated Sun Tzu could be expanded to cover more areas. For instance, military texts quoting Sun Tzu from other English-speaking countries such as Australia, Canada, and New Zealand need to be added. Texts quoting English translation of Sun Tzu from non-English-speaking countries could also be included. If more texts from other disciplines, such as business management, are included, the effect of translation on knowledge transfer from military discipline to other disciplines can be investigated.

Thirdly, a comparative investigation between the translation of ancient Chinese military classics and modern Chinese military masterpieces into English could offer us fresh perspectives. Similarly, if we compare the translation of *The Art of War* into Western countries with the translation of *On War* into China, there would be even more interesting findings on the role of translation in the exchange of strategic culture.

Last but not least, since this study is interdisciplinary, covering areas of translation, military and culture, future work could be done through collaboration among scholars of different academic backgrounds. Such cooperation would clearly lead to more fresh and exciting findings.

Appendices

Appendix 1: A list of English translations of *The Art of War* (in chronological order of the first version of each translator)

1　Caltop, Everard F. 1905. *Sonshi: The Chinese Military Classic*. Tokyo: Sanseido.

　　Calthrop, Everard F. 1908. *The Book of War: The Military Classic of Far East*. London: John Murray.

2　Giles, Lionel. 1910. *Sun Tzu on The Art of War: The Oldest Military Treatise in the World (Translated from the Chinese with Introduction and Critical Notes by Lionel Giles)*. London: Luzac.

　　吉利(Lionel Giles)英译 陶希圣(Hsi-sheng T'ao)校订. 1954, 1964. 孙子兵法. 台北: 全民.

　　吉尔士(Lionel Giles)英译 时超语译. 1972, 1977. 孙子兵法(中英对照/文白对照). 台北: 文致出版社.

　　Giles, Lionel. 1983. *The Art of War: Sun Tzu (Edited and Foreword by James Clavell)*. New York: Delacorte Press.

　　Giles, Lionel. 2004. *The Art of War (Edited and Introduction by Dallas Galvin)*. New York: Barnes & Noble Classics.

　　Giles, Lionel. 2007. *Sun Tzu's The Art of War (Bilingual Edition Complete Chinese and English, with a New Foreword by John Minford)*. North Clarendon, VT: Tuttle Publishing.

　　Giles, Lionel. 2012. *The Illustrated Art of War: Sun Tzu (Edited by Andrew Forbes and David Henley)*. Chiang Mai: Cognoscenti Books.

3　李浴日. 1938. 孙子兵法之综合研究. 商务印书馆.

4　Machell-Cox, E. 1943. *Principles of War by Sun Tzu*. Ceylon: Royal Air Force.

5　Sadler, Arthur L. 1944. *Three Military Classics of China: The Art of War of Sun Tzu, The Precepts of War by Sima Rangju, Wu Zi on The Art of War*. Sydney: Australasian Medical Publishing Company, Ltd.

　　Sadler, Arthur L. 2009/2011. *Chinese Martial Code: The Art of War of Sun Tzu, The Precepts of War by Sima Rangju, Wu Zi on The Art of War (Bilingual Edition, Annotated English Translations with Complete Chinese text, Translations by Arthur Lindsay Sadler, Foreword and Annotations by Edwin Lowe)*. North Clarendon, VT: Tuttle Publishing.

6　郑麐编译（Cheng Lin）. 1945. 孙子兵法（*The Art of War: Military Manual Written circa. B.C. 510, The Original Chinese Text Appended,*

Translated with an Introduction by Cheng Lin). 重庆（Chungking): 中国
辞典馆（World Encyclopedia Institute).

郑麐编译. 1946. 孙子兵法（附英文译文). 上海: 世界书局.

郑麐编译. 1953, 1972, 1975. *孙子兵法（杨家洛主编，中英对照，白
话语译)*. 台北: 世界书局.

7   戴冕伦英译（Tai Mien-leng). 1954. 孙子兵法(T*he Art of War*, Advocated
and Written by Sun Tzu, Translated by Tai Mien-leng). 台北：军用图书
公司.

8   Griffith, Samuel B. 1963. *The Art of War* (forward by Liddell Hart).
New York: Oxford University Press.

Griffith, Samuel B. 2005. *The Illustrated Art of War (Forward by Liddell
Hart)*. New York: Oxford University Press.

9   Tang, Zichang. 1969. *Principles of Conflict: Recompilation and New
English Translation with an Annotation on Sun Zi's Art of War*. San
Rafael, CA: T. C. Press.

唐子长编译. 1971. *孙子重编 (Principles of Conflict, 唐子长编译, 中英
对照本)*. 香港: 南天书业公司.

10 葛振先著 周以鸿图. 1973. *孙子兵法中英文详解*. 台北: 正中书局.

11 Chen, A., and Chen, C. 1982. *Sun Tzu. The Art of War: A Treatise on
Chinese Military Science Compiled About 500 B.C.* Singapore: Graham
Brash (PTE) Ltd.

12 Leong, Wen Kam. 1986. *Chinese Military Classic: The Art of War (Edited
and Illustrated by Tsai Chih Chung and Translated by Leong Weng Kam)*.
Singapore: China Books & Periodicals.

Leong, Wen Kam. 1991. *Chinese Military Classic: The Art of War
(Edited and Illustrated by Tsai Chih Chung)*. Singapore: Asiapac.

13 Yuan, Shibing. 1987/1990. *Sun Tzu's Art of War: The Modern Chinese
Interpretation (Annotated by Tao Hanzhang; Translated by Yuan Shibing)*.
New York: Sterling Publishing Company Incorporated.

Yuan, Shibing. 2007. *Sun Tzu's Art of War: The Modern Chinese
Interpretation (Annotated by Hanzhang Tao)*. New York: Sterling
Publishing Company Incorporated.

14 Wing, R. L. 1988. *The Art of Strategy: A New Translation of Sun Tzu's
Classic The Art of War*. 1st ed. New York: Doubleday.

15 潘嘉玢, 刘瑞祥译. 1990. *孙子校释 (吴九龙主编，中，英，法，俄，
日，意5种文字，第1版)*. 北京：军事科学出版社.

16 Cleary, Thomas. 1991. *The Art of War Sun Tzu*. Boston and
New York: Shambhala Publications.

Cleary, Thomas. 1998. *The Illustrated Art of War*. 1st ed. Boston and
Shaftesbury: Shambhala Publications.

17 罗顺德编译. 1991. *孙子兵法 (中英对照本)*. 台北: 黎明文化事业股份有
限公司.

18 刘海明译. 1992. *孙子名言选译（田昌五编译；刘海明英译)*. 济南: 齐鲁
书社.

19 蒲元明译. 1992. *白话英译孙子兵法（黄葵编著；蒲元明英译）*. 重庆：重庆出版社.

20 Ames, Roger T. 1993. *The Art of Warfare: Classics of Ancient China (The First English Translation Incorporating the Recently Discovered Yin-chueh-shan Texts, with an Introduction and Commentary)*. 1st ed. New York: Ballantine Books.

21 Sawyer, Ralph D. 1993. *The Seven Military Classics of Ancient China (with a Commentary)*. Boulder, CO: Westview Press.

    Sawyer, Ralph D. 1994. *The Art of War by Sun Tzu (with Introduction and Commentary, with Collaboration of Mei-chun Lee Sawyer)*. New York: Barnes & Noble Books.

    Sawyer, Ralph D. 1996. *The Complete Art of War: The Art of War by Sun Tzu and The Art of War by Sun Bin (with Historical Introduction and Commentary)*. Boulder, CO: Westview Press.

    Sawyer, Ralph D. 2005. *The Essential Art of War (Translated and Interpreted by Ralph D. Sawyer)*. 1st ed. New York: Basic Books.

22 Bruya, Brian. 1994. *Sunzi Speaks: The Art of War (Adapted and Illustrated by Tsai, Chih Chung)*. New York: Anchor Books.

    Bruya, Brian. 2005. *孙子说: 兵学的先知（蔡志忠绘）*. 北京: 现代出版社.

    Bruya, Brian. 2018. *The Art of War (Adapted and illustrated by Tsai, Chih Chung, Foreword by Lawrence Freedman)*. Princeton and Oxford: Princeton University Press.

23 罗志野. 1995. *100 Sun Tzu's The Art of War 孙子兵法一百则*. 台北: 台湾商务印书馆.

    罗志野译. 1996. *孙子兵法一百则（罗志野译）*. 北京：中国对外翻译出版公司；商务印书馆（香港）有限公司.

24 Zhang, Huimin. 1995. *Sun Zi: The Art of War with Commentaries (Annotated by Xie Guoliang)*. Beijing: Panda Books.

    张惠民译. 1995. *孙子兵法与评述（谢国良评注）*. 北京: 中国文学出版社.

25 林戊荪译. 1995. *孙子兵法•孙膑兵法（吴如嵩，吴显林校释）*. 北京: 人民中国出版社.

    Lin, Wusun. 2003. *The Art of War*. 1st ed. San Francisco: Long River Press.

26 Kaufman, Stephen F. 1996. *The Art of War: The Definitive Interpretation of Sun Tzu's Classic Book of Strategy*. Clarendon, VT: Tuttle Publishing.

27 Zhong, Qin. 1996. *The Essentials of War the Masterpiece of a Strategist in Ancient China (Bilingual Edition)*. Beijing: New World Press.

28 Chen, Bingfu, and M.W.L. Chan. 1998. *Sunzi on The Art of War and Its General Application to Business*. Shanghai: Fudan University Press.

29 Gagliardi, Gary. 1999. *The Art of War: In Sun Tzu's Own Words*. Seattle: Clearbridge Publishing.

    Gagliardi, Gary. 2003. *The Art of War Plus the Ancient Chinese Revealed*. Hillsborough, WA: Clearbridge Publishing.

30 Tan, Han H. 2001. *Sun Tzu's The Art of War*. Queensland: H. H. Tan Medical PL Ltd.

31 The Denma Translation Group. 2001. *The Art of War: The Denma Translation*. Boston: Shambhala Publications.
    The Denma Translation Group. 2002. *The Art of War (Translation, Essays and Commentary by the Denma Translation Group)*. Boston: Shambhala.

32 Minford, John. 2002. *The Art of War (with an Introduction and Commentary by John Minford)*. New York: Viking.

33 Tarver, D. E. 2002. *The Art of War Sun Tzu's Classic in Plain English with Sun Pin's The Art of Warfare*. Lincoln, NE: Writers Club Press.

34 龚理曾、杨爱文译. 2002. *孙子的故事 英文版 (曹尧德，曹笑梅著)*. 北京：外文出版社.

35 Cantrell, Robert L. 2003/2004. *Understanding Sun Tzu on the Art of War*. Arlington, TX: Center for Advantage Company.

36 Chohan, Chou-wing, and Abe Bellenteen. 2003. *The Art of War: The Cornerstone of Chinese Strategy (edited by Brant, Rosemary)*. Hod Hasharon, Israel: Astrolog Publishing House.

37 Wee, Chow H. 2003. *Sun Zi Art of War: An Illustrated Translation with Asian Perspectives and Insights*. Singapore: Pearson Prentice Hall.

38 Richter, Gregory C. 2004. *Sūn Zǐ Bīng Fǎ Sun Zi's Art of War (Pinyin Transcription, Gloss, and English Translation by Gregory C. Richter)*. Kirksville, MO: Truman State University.

39 Sui, Yun. 2004. *Sunzi's Art of War: World's Most Famous Military Classic (Illustrated by Wang Xuanming)*. 10th ed. Singapore: Asiapac.

40 Colwell, Eric. 2005. *Master Sun's Art of War*. Ypsilanti, MI: Eastern Michigan University.

41 王之光译. 2005. 孙子兵法（中英文丝绸彩印珍藏本). 杭州：西泠印社出版社.

42 Huang, J. H. 2008. *The Art of War: The New Translation*. New York: Harper Perennial Modern Classics.

43 Mair, Victor H. 2008. *The Art of War: Sun Zi's Military Methods*. New York: Columbia University Press.

44 Thomas Huynh. 2008. *The Art of War: Spirituality for Conflict Annotated & Explained*. Woodstock, VT. Skylight Paths Publishing.

45 王琴、姜防宸译. 2008. *孙子精华录（英汉对照，蔡希勤编注)*. 香港: 瀚林苑出版社.

46 李庆山译注. 2008. *图读孙子兵法*. 上海：上海辞书出版社.

47 Makah, Jonathan, and Marques Jalil. 2009. *The Art of War Sun Tzu*. Newark, NJ: GAU Publishing.

48 Patrick E. Moran. 2010. *Master Sun's Art of War: A Classical Text for Modern Martial Artist (Translation and Commentary by Patric Edwin Moran)*. Patrick Edwin Moran.

49 宋德利译. 2010. *孙子兵法（汉英对照)*. 哈尔滨：哈尔滨出版社.

50 Ivanhoe, Philip J. 2011. *Master Sun's Art of War (Translated with Introduction by Philip J. Ivanhoe)*. Indianapolis: Hackett Publishing Co.

51 Trapp, James. 2011. *The Art of War: A New Translation*. London: Amber Books.

52 Clements, Jonathan. 2012. *The Art of War: A New Translation*. London: Constable.

53 Jones, David G. 2012. *The School of Sun Tzu: Winning Empires without War*. Bloomington, IN: Iuniverse, Inc.

54 Roman, Kelly. 2012. *The Art of War: A Graphic Novel*. 1st ed. Translated by Michael DeWeese. New York: Harper Perennial.

55 肖刚. 2013. 领导科学第一书 *兵法 汉英对照*. 北京: 中国经济出版社.

56 金精译. 2013. *孙子兵法精读 （汉英对照）*. 西安: 西安交通大学出版社.

57 McNeilly, Mark, and Mark R. McNeilly. 2014. *Sun Tzu and the Art of Modern Warfare*. New York: Oxford University Press.

58 贺少华译. 2014. 孙子兵法与现代战争（贺少华编著）. 长沙: 中南大学出版社.

59 远翔、李京著；王壹晨、王国振译. 2014. *兵圣一孙子（英文版）*. 北京: 五洲传播出版社.

60 Hagy, Jessica. 2015. *The Art of War Visualized: The Sun Tzu Classic in Charts and Graphs*. New York: Workman Publishing Company.

61 黎金飞释析, 沈菲译, 尹红等绘. 2016. *孙子兵法（汉英对照）*. 桂林: 广西师范大学出版社.

62 任力、魏鸿、高润浩著, 林戊茹、魏怡译. 2017. 名家讲《孙子兵法》（英文版）. 北京: 五洲传播出版社.

63 Harris, Peter. 2018. *The Art of War*. Vol. 385 of *Everyman's Library*. New York: Alfred A. Knopf.

64 Nylan, Michael. 2020. *The Art of War: A New Translation*. New York: W. W. Norton & Company.

Appendix 2: A sample list of high-frequency Chinese words in *The Art of War*

Rank	Word	Frequency	Percentage
1	之	317	4.9617
2	者	194	3.0365
3	也	186	2.9113
4	而	176	2.7547
5	则	84	1.3148
6	其	82	1.2835
7	不	79	1.2365
8	敌	54	0.8452
9	以	54	0.8452
10	于	50	0.7826
11	地	49	0.7669
12	有	45	0.7043
13	战	45	0.7043
14	曰	41	0.6417
15	必	38	0.5948
16	无	38	0.5948
17	胜	37	0.5791
18	利	36	0.5635
19	人	36	0.5635
20	兵	33	0.5165
21	将	33	0.5165
22	知	33	0.5165
23	为	32	0.5009
24	所	31	0.4852
25	此	30	0.4696
26	故	30	0.4696
27	可	29	0.4539
28	众	28	0.4383
29	间	27	0.4226
30	吾	26	0.4069
31	不能	25	0.3913
32	如	24	0.3756
33	不可	23	0.36
34	用	23	0.36
35	军	22	0.3443
36	可以	22	0.3443
37	形	22	0.3443

Rank	Word	Frequency	Percentage
38	击	20	0.313
39	攻	19	0.2974
40	使	19	0.2974
41	用兵	19	0.2974
42	得	17	0.2661
43	动	17	0.2661
44	五	17	0.2661
45	与	17	0.2661
46	至	17	0.2661
47	不知	16	0.2504
48	法	16	0.2504
49	能	16	0.2504
50	生	16	0.2504

Appendix 3: Rhetorical figures in *The Art of War*

Rhetorical figures	Occurrences
Antithesis	92
Parallelism	77
Metaphor	28
Simile	19
Anaphora	13
Hyperbole	12
Repetition	6
Antimetabole	6
Contrast	5
Allusion	3
Anticlimax	2
Comparison	2
Metonymy	2
Anadiplosis	2
Paradox	2
Synecdoche	2
Total	273

Appendix 4: Similes and metaphors in the ST core text

V	ST	Figure of Speech
1	蚁附之	simile
2	胜兵若以镒称铢,败兵若以铢称镒	simile
3	胜者之战民也，若决积水于千仞之溪者，形也。	simile
4	兵之所加，如以碬投卵者，虚实是也。	simile
5	奇正相生，如环之无端，孰能穷之？	simile
6	势如张弩	simile
7	节如发机	simile
8	任势者，其战人也，如转木石。木石之性，安则静，危则动，方则止，圆则行。	simile
9	故善战人之势，如转圆石于千仞之山者，势也。	simile
10	夫兵形象水，水之形避高而趋下；兵之形，避实而击虚。水因地而制流，兵应敌而制胜。故兵无常势，水无常形。	simile
11	其疾如风	simile
12	其徐如林	simile
13	侵掠如火	simile
14	不动如山	simile
15	难知如阴	simile
16	动如雷震	simile
17	故善用兵者，譬如率然。率然者，常山之蛇也。击其首则尾至，击其尾则首至，击其中则首尾俱至。敢问："兵可使如率然乎？"曰："可。"	simile
18	夫吴人与越人相恶也，当其同舟而济，遇风，其相救也，如左右手。	simile
19	焚舟破釜，若驱群羊；驱而往，驱而来，莫知所之。	simile
21	始如处女，敌人开户	simile
22	后如脱兔，敌不及拒	simile
20	践墨随敌，以决战事。	metaphor
23	激水之疾，至于漂石者，势也；	metaphor
24	鸷鸟之疾，至于毁折者，节也。	metaphor
25	悬权而动。	metaphor
26	饵兵勿食	metaphor
27	故进不求名，退不避罪，惟人是保，而利合于主，国之宝也。	metaphor

Appendix 5: Quotes on the flyleaf of *Strategy: Indirect Approach* (Hart 1954)

"All warfare is based on deception. Hence, when able to attack, we must seem unable; when using our forces, we must seem inactive; when we are near, we must make the enemy believe that we are away; when far away, we must make him believe we are near. Hold out baits to entice the enemy. Feign disorder, and crush him."

"There is no instance of a country having been benefited from prolonged warfare."

"It is only one who is thoroughly acquainted with the evils of war that can thoroughly understand the profitable way of carrying it on."

"Supreme excellence consists in breaking the enemy's resistance without fighting.

Thus the highest form of generalship is to baulk the enemy's plans; the next best is to prevent the junction of the enemy's forces; the next in order is to attack the enemy's army in the field; the worst policy of all is to besiege walled cities."

"In all fighting, the direct method may be used for joining battle, but indirect methods will be needed in order to secure victory."

"Appear at points which the enemy must hasten to defend, march swiftly to places where you are not expected."

"You may advance and be absolutely irresistible, if you make for the enemy's weak points; you may retire and be safe from pursuit if your movements are more rapid than those of the enemy."

"All men can see these tactics whereby I conquer, but what none can see is the strategy out of which victory is evolved."

"Military tactics are like unto water; for water in its natural course runs away from high places and hastens downwards. So in war, the way to avoid what is strong is to strike what is weak."

"Water shapes its course according to the ground over which it flows; the soldier works out his victory in relation to the foe whom he is facing."

"Thus, to take, a long circuitous route, after enticing the enemy out of the way, and though starting after him, to contrive to reach the goal before him, shows knowledge of the artifice of deviation."

"He will conquer who has learnt the artifice of deviation. Such is the art of manoeuvring."

"To refrain from intercepting an enemy whose banners are in perfect order, to refrain from attacking an army drawn up in calm and confident array—this is the art of studying circumstances."

"When you surround an army leave an outlet free. Do not press a desperate foe too hard."

"Rapidity is the essence of war; take advantage of the enemy's unreadiness, make your way by unexpected routes, and attack unguarded spots."

SUN TZU, The Art of War—500 B.C.

"The most complete and happy victory is this: to compel one's enemy to give up his purpose, while suffering no harm oneself."

BELISARIUS

"By indirections find directions out."

SHAKESPEARE, Hamlet, Act II, Scene I

"The whole art of war consists in a well-reasoned and extremely circumspect defensive, followed by rapid and audacious attack."

NAPOLEON

"All military action is permeated by intelligent forces and their effects."

CLAUSEWITZ

"A clever military leader will succeed in many cases in choosing defensive positions of such an offensive nature from the strategic point of view that the enemy is compelled to attack us in them."

MOLTKE

"Gallant fellows, these soldiers; they always go for the thickest place in the fence."

ADMIRAL DE ROBECK—

watching the Gallipoli landing, 25th April 1915.

Appendix 6: Sample word clusters beginning with "communist" (first 40 items)

Serial		
1	451	communist party
2	89	communist forces
3	65	communist China
4	44	communist leaders
5	40	communist insurgency
6	37	communist military
7	37	communist parties
8	36	communist movement
9	36	communist world
10	33	communist revolution
11	32	communist revolutionary
12	29	communist regime
13	26	communist ideology
14	25	communist state
15	24	communist leadership
16	22	communist international
17	21	communist bloc
18	20	communist control
19	20	communist strategy
20	19	communist aggression
21	19	communist and
22	19	communist army
23	19	communist states
24	17	communist countries
25	17	communist leader
26	17	communist regimes
27	17	communist victory
28	16	communist government
29	16	communist rule
30	15	communist giants
31	15	communist guerrilla
32	14	communist base
33	14	communist troops
34	13	communist power
35	12	communist armies
36	12	communist Chinese
37	11	communist force
38	11	communist soldiers
39	11	communist-led
40	10	communist conquest

Index